皮尤 + 斯卡帕
Report / 2005

Acknowledgements

This publication has been made possible with the help and cooperation of many individuals and institutions. Grateful acknowledgement is made to Pugh + Scarpa, for its inspiring work and for its kind support in the preparation of this book on Pugh + Scarpa for the AADCU Book Series of Contemporary Architects Studio Report In The United States.

©Pugh + Scarpa
©All rights reserved. No part of this publication may be reproduced, stored in a retrieval system or transmitted in any form or by means, electronic, mechanical, photocopying, recording or otherwise, without the permission of AADCU.

Office of Publications:
United Asia Art & Design Cooperation
www.aadcu.org
info@aadcu.org

Project Director:
Bruce Q. Lan

Coordinator:
Robin Luo

Edited and published by:
Beijing Office, United Asia Art & Design Cooperation
bj-info@aadcu.org

China Architecture & Building Press
www.china-abp.com.cn

In Collaboration with:
Pugh+Scarpa
www.pugh-scarpa.com

d-Lab & International Architecture Research

School of Architecture, Central Academy of Fine Arts

Curator/Editor in Chief:
Bruce Q. Lan

Book Design:
Design studio/AADCU

ISBN: 7-112-07385-5

©本书所有内容均由原著作权人授权美国亚洲艺术与设计协作联盟编辑出版，并仅限于本丛书使用。任何个人和团体不得以任何形式翻录。

出版事务处：
亚洲艺术与设计协作联盟／美国
www.aadcu.org
info@aadcu.org

编辑与出版：
亚洲艺术与设计协作联盟／美国
bj-info@aadcu.org

中国建筑工业出版社／北京
www.china-abp.com.cn

协同编辑：
皮尤＋斯卡帕
www.pugh-scarpa.com

国际建筑研究与设计中心／美国

中央美术学院建筑学院／北京

主编：
蓝青

协调人：
洛宾·罗，斯坦福大学

书籍设计：
设计工作室／AADCU

SERIES OF CONTEMPORARY ARCHITECTS STUDIO REPORT IN THE UNITED STATES

Pugh + Scarpa

Contents

引言，7页
项目研究
—科罗拉多法院公寓，23页
—COop剪辑工作室，41页
—生命之源项目，59页
—迪瓦工作室，67页
—橘林住宅，77页
—北点住宅项目，85页
—Jigsaw公司，95页
—太阳能伞住宅，109页
—圣莫尼卡学院学生服务中心，117页
—伯格莫特艺术家阁楼，125页
—韦尔格兰特住宅，139页
—音乐电影制作公司，149页
—XAP项目，159页
—Reactor电影工作室项目，171页
建筑师年表，181页

Introduction with Article Review, Page 7. Project Survey Colorado Court Page 23. COop Editorial Page 41. Nascent Terrain Page 59. Diva Page 67. Orange Grove Page 77. North Point Page 85. Jigsaw Page 95. Solar Umbrella Page 109. SM College Student Service Center Page 117. Bergamot Loft Page 125. Vail Grant House Page 139. The Firm Page 149. XAP Page 159. Reactor Films Page 171. Chronology Page 181 & Bibliography

Pugh + Scarpa

Contents

Introduction with Article Review, Page 7.

Project Survey Colorado Court COop Editorial Nascent Terrain Diva Orange Grove North Point Jigsaw Solar Umbrella SM College Student Service Center Bergamot Loft Vail Grant House The Firm XAP Reactor Films Chronology

Introduction with Article Review

Interview with Lawrence Scarpa

V5: Pugh + Scarpa has developed a fast response design studio which works closely with a wide range of trades. How does this work?

LS: Each project is different so the team and relationships change. In the case of our project "Inside Out", a renovation of a building built in the 1960's, the owner wanted to convert it into a place where they could do off-line editing. The first thing that really struck me was how the courtyards were buried internally within the offices, and we really wanted to make them part of the public space. I worked from the conditions that presented themselves. The found conditions were inspiring to me and something that has always been of great interest. A lot of the paneling and the exterior doors on the exterior were reused in the interior. We wound up reusing quite a bit of the materials that were already there by virtue of stripping and reconfiguring them.

V5: Did your team design all the built in furniture?

LS: Yes. We were using a lot of the materials that we found on the site for this as well.

V5: That requires a level of "hands on" evaluations as you go through a project.

LS: Absolutely. A lot of times we wind up changing our minds in the middle of things and having to deal with decisions that we have.

V5: Does that scare the client?

LS: Initially, yes. But we have gotten into a position where they are more comfortable with the interiors that we do by virtue of our past portfolio. But in the beginning it was very difficult.

V5: You are working with a very rich palette of materials; existing woods, metals, plaster and drywall. Is that hard to control?

LS: In this case, the materials stemmed from the existing wood in the office, so we used that as a base for our palette. The steel that you see is actually something that we developed with the fabricator, who is incidentally a great craftsman. We did probably a hundred samples of steel and this is just cold rolled, eighteen-gauge steel, we sandblasted it then started rubbing it with gun blue and copper.

V5: Did you make your own finish recipe?

LS: Yes, we developed our own recipe to come up with the kind of finish that is there. So what really interests me and has taken me a long time to learn is what you can find in simplicity. I have become less interested in form-making and more interested in space shaping and the experience of it. In fact, I taught a studio this past semester at Otis College with my wife, Angela Brooks, where we did an interiors project. Unlike most student projects, we did not concern ourselves with form making at all, but did material exercises. We looked at how we could turn the horizontal grains, how the verticals changed with subtle patterns, things you also see in steel as well as in wood.

By Mark Dillon for Volume 5

问：皮尤+斯卡帕设计事务所已经发展成一个能做出快速反应的设计工作室，并和不同的行业进行着广泛而紧密的合作。它的具体运作情况是怎样的？
答：每个项目都有不同之处，所以团组以及它们之间的关系都会有所变化。在我们的项目"Inside Out"———一个20世纪60年代建筑的改建例子中，业主希望能把它变成一个可以适应做离线编辑的场所。这个项目给我印象最深的事情是如何在办公区内部设计一座庭院，当时我们真地想把它作为公共空间的一部分。我从现场的实际情况出发，发现现场非常令我振奋而且总是那么有趣。原有建筑外部的很多嵌板和门都在内部得以重新使用。我们通过剥离和改装利用了很多现有材料。

问：你的团队负责设计所有的家具吗？
答：是的，我们为此也利用了很多场地内原有的材料。

问：你们在项目过程中需要一定程度的实际操作来应对变化吗？
答：绝对是这样的。有很多次我们都落得在项目进展期间改变方案设想，并且不得不因此采取相应的对策。

问：这会吓着你们的客户吗？
答：开始的时候会这样。但是我们会通过以往的经验让他们对这些内部设计感到放心。当然，开始的时候还是很困难的。

问：你们使用了各种各样的材料：现有的木材、金属、塑料和干墙。控制起来困难吗？
答：在这个项目中，材料来源于办公室中现存的木材，我们以此为基础。你看到的钢材是我们和一个出色的工艺师共同开发的。我们可能开发了100多个钢材样本，这是冷轧的18号（直径为18）钢，我们在上面喷了沙，然后用金属染黑剂和铜进行了处理和打磨。

问：你们有自己的涂料配方吗？
答：是的。我们依据实际需要开发自己的解决方案。真正使我感兴趣并且让我花很长时间来学习的是简约的要素。我对形式已不那么感兴趣了，相反对空间塑造和体验更感兴趣。上个学期，我和妻子安吉拉·布鲁克斯在奥特斯艺术与设计学院教授设计课程，在那里我们完成了一个室内设计项目。与其他学生的方案不同，我们根本没有考虑形式创造，而是着重于材料实践。我们看到了横向的粒子是如何转变的，竖向的粒子是

劳伦斯·斯卡帕访谈

马克·迪龙

V5: In your design process, how do you visualize these factors in order to bring that kind of decision-making information to the project?

LS: With this particular project we did do all drawings. Generally we try to do it differently for each project so that we get a fresh perspective each time. For example, I find that working in the same method over and over again, whether it is doing drawings or making models, will have a certain predictability of what the outcome is. We try to say, 'let's not do it the way we did it in the last project and let's try something completely different.' What that does is it makes us see things that we would not normally see.

V5: There is clarity of all the materials.

LS: Yes. That is an interest that repeats for us over and over, this kind of idea of surface wrapping or making space through surfaces that create a negative space. You can see things that penetrate through such as ceiling planes that turn into wall planes and connect to the floor planes. The materials are important. How can you put a material on the floor and wall that will work in both cases?

V5: It adds a level dynamic quality to the space, there is not a bottom, middle or top. It is not a classic language.

LS: Right. Which is what I am interested in and intrigued with. What we try to do is to make each piece extremely readable and the whole project will in some ways become more abstracted through readable pieces.

One of the things that really inspired me was something Robert Venturi said, "Familiar things seen in an unfamiliar way become both perceptually old and new at the same time." So when you are able to take things, for example, this (shipping) container, it already has a history or richness already built in it. So to weave in some piece, so to speak, of history or something that someone relates to, it already has a sort of richness to it. Then we transform it into something else.
In the case of the (shipping) container, it has a incredible history of housing of goods, being moved from east to west, it has it's own baggage attached to it before you even touch it. Then when you alter it you develop an even richer meaning and story to it.

V5: It is very collage like in that the first reading of it is not a (shipping) container at all, it has been manipulated to a point that it is clearly something quite different.

LS: We are not interested in just taking down things and throwing them into the space.

V5: You have authorship of the design.

LS: Yes. We treat the things pretty seriously and one thing that we have become fairly good at as an office is the pragmatic issues. A lot of architects have a very poor sigma attached to them regarding cost and detailing. When we design a conference room, we make it acoustically sound. We try to

怎样变成微妙的图案，这些在钢和木材中都可以看见。

问：在你的设计过程中，视觉化因素是如何引入项目的决策性信息中的？
答：在重点项目里，我们的确做了所有部分的绘图工作。通常我们会努力使每一个项目的绘图方法都与众不同，这样的话，我们可以时刻更新自己的眼界。比如我发现无论是绘图还是制模，长期使用一种相同方式的结果都是可以预见的。我想说："让我们不要重复以前项目的方法，让我们尝试完全不同的东西吧。"这使我们能够看到通常看不到的事物。

问：所有材料的使用都很明确吧？
答：是的，表层包裹或者通过表层包裹制造一个负空间的概念是不断吸引我们的一个原因。你可以通过材料看到顶棚、墙面和地面的结合。材料很重要，你怎样才能找到一个同时适用于地板和墙壁的材料呢？这很有挑战性。

问：这给空间加入了一定程度的活力。不存在底部、中间或者顶端的界限。传统的语言是不能诠释它的。
答：对。我所感兴趣和迷恋的就是这一点。我们努力做的就是使每一个元素都能被最大程度地理解，并且通过这些易懂的元素使整个项目从某种程度上变得更加抽象。罗伯特·文丘里(Robert Venturi)曾经说过一段给我带来极大的灵感的话："一个原本为我们所熟悉的事物，经由新的方法改造过后，会带给人们新老并存的感觉。"所以，当你在拿起一样东西的时候，比如，看这个集装箱，要知道它已经拥有一段历史或者更多的含义。那么就把这些相关的历史、相关的人物都编织起来、表达出来吧。这样一来，我们就把它转化成为不同的事物。这个集装箱拥有神奇的货运历史，它从东方来到西方，在你触摸它以前，它曾装有自己的内容。然而当你改变它的时候，你也许会给予它更丰富的意义和故事。

问：刚看上去就像抽象拼贴画，一点也不像个集装箱。它已经被巧妙地处理成完全不同的东西了。
答：要知道，我们对卸下东西，然后把它们扔到空间中这种做法并不感兴趣。

问：你们拥有设计专利吗？
答：有。我们对此非常严肃。作为一个工作团体，我们解决实际问题的能力很强。很多

treat the budget and program quite seriously when we deal with the project. We take a lot of risks and many times we have to make amends for that, yet we treat it seriously. Gramercy Group Homes was a project that we did with some SCI-Arc students and this was a sixteen-unit rehab for single, teenage mothers, a non-profit group in the Crenshaw district of Los Angeles. We did a one-week study and built this project. This was actually one of my student's ideas, Wendy Bone. My client provided them with a space, since they had no money. SCI-Arc kicked in a few dollars and we were funded. The students built the furniture for the mothers as well. You can see the back of the structure and one of the interiors, a very small unit, about three hundred square feet each. Jackson, who works in my office, made all the furniture and we did these tables with a storage box on the side, it is a very large table so that they can do homework and dine. It has a little storage component with a door that flips down and this is the little kid's worktable. Of course, you get a lot of interest at first, but it wound up being a small, dedicated group of people that really did the work. This non-profit group built all this very inexpensively. I am quite proud of this work.

V5: Did these projects lead to the new residential projects that you are now working on?

LS: Partly. We have done work with the schools here where we have actually worked with the elementary school kids and art teachers where we actually have made tiles and plaques. Now they hang in the school hallways. Another collaboration-based project was the electric vehicle charging station; we did that in conjunction with Tony Louie, and John Ingersol. We wrote a grant through a state assembly bill for clean air project securing funding to build this project. Then we went to the City of Santa Monica, saying we have money to build this project, can you provide a site for us, and they did. So we are active in trying to create projects as well as waiting for projects to come to us. The kinds of projects we try to create are one that involve the community and reach out to other people. We did a housing complex project in South Central Los Angeles with Roger Sherman, which was an invited competition, we didn't win. It consisted of forty units of housing, a mixed-use project and we put together a design-build team, which included a non-profit developer and contractors, and was sponsored by First Interstate Bank.

V5: How many other participants where there in the competition?

LS: It started out open, they then pared it down to ten and then down to three.

V5: Is the project going forward?

LS: Yes, it is under construction now. Dan Solomon out of San Francisco won the competition. But

建筑师对于成本和细节问题不是很关注。当我们在设计一个会议室的时候，我们会使它的声学效果非常合理。我们非常认真地对待项目的预算和计划。为此我们冒了很大的风险，花了很多功夫进行修改工作。格拉默西家园（Gramercy Group Home）是一个我们和南加利福尼亚建筑学院学生合作的改建项目，它是洛杉矶克伦肖地区一个非赢利组织所属的专门为未成年的单身母亲设计的16个单元的住宅。我们用了一个星期进行调查，之后完成了这个项目。而项目的最初设想来自我的一个学生温迪·博恩(Wendy Bone)。因为这些未成年的单身母亲没有钱，我的业主给她们提供住处。南加利福尼亚建筑学院和我的事务所也给予了资助。学生们还为这些年轻的母亲设计了家具。你可以看到结构的后面和其中一个室内设计，那是一种很小的单元，每一个面积大概300平方英尺。我的一个同事杰克逊(Jackson)设计了所有的家具，桌子很大，可以用来做功课和用餐。这些桌子的一侧配有可以拉出的带门的小储存箱可作孩子们的作业桌。当然，你在一开始会非常感兴趣，但是，真正让我们感动的是这群年轻的无私奉献的学生。这个为非赢利组织建造的建筑成本低廉。我们为此感到非常骄傲。

问：是这些项目的成功给你们正在做的住宅项目带来了新的启示吗？

答：部分是。我们和这里的学校，主要是和初级学校的孩子和艺术教师进行了合作。另外一个合作项目是电动汽车充电站，那次的合作者是托尼·路易（Tony Louie）和约翰·因吉索（John Ingersol）。我们通过为空气净化项目制定的州议会提案而获得了足够的资金。然后，我们去了圣莫尼卡市，向他们表示我们已经拥有了足够的资金，希望他们能够给我们提供场地，最终我们获得了该市的支持。我们一边努力完成手中的项目，一边等待新的项目机会。我们一直在努力创造更多的关于社区一类的项目。我们曾经在洛杉矶中南部与罗杰·谢尔曼（Roger Sherman）合作一个复合型住房竞标项目，但没有中标。它是一个包括40个单元的混合功能住宅，我们的设计团队包括一个非赢利开发商和承包人，赞助商是第一州际银行。

问：总共有多少竞标者？
答：开始阶段不限竞标者数量，后来淘汰到10名，最后剩下3名。

问：这个项目还在继续吗？

our strategy was our belief in doing the right thing, making it home ownership, as opposed to rentals. We provided every single home with a private garden, so rather than it be condominiums with a balcony, on all levels every unit had a garden. That in turn made our project more expensive and we needed entitlements for it to get done, so it was a much more difficult undertaking than Dan Solomon's project. I think that was part of the reason why we didn't win but we obviously thought ours was a much better project.

V5: Did First Interstate Bank pay for your development costs?

LS: We did receive some money to prepare a design, which is unlike most competitions. It just means we lost that much less. (laughs)

V5: We just did an interview with Yo Hakomori and the team (HPST) that worked on the Beijing competition, and they spoke about how expensive in time and material it is. It is a fairly extraordinary undertaking. Do you find your studio doing competitions often?

LS: We do them, but try to be selective about them. We pick competitions based on what our real interests are. I think that helps our studio.

We talked about how the contractor is part of our team and we generally do things without bidding them, so it's negotiated in the budget, and we start construction often within the first week of design. We really work as a team. Our project "Click3xLA" started with the idea of a translucent wall. All the offices are clad in translucent corrugated fiberglass. We wanted it to glow. Many times you will go into large film companies and their offices are dismal, it's like they turn their back on them to make a kind of pretty form. They are dark, no light in them and are a terrible work environment. It's important for us to look at the offices as an important work place and getting natural light into it was part of it. So we were looking for ways to maintain privacy, which was very important to them, but also get light. The wall acts as a translucent filter, which also illuminates the space when the lights are on.

V5: Is the carpet of your design as well?

LS: Yes, the carpet, coffee table and reception desk we designed as well. This was done in a similar fashion where we take common carpets and stitch them together into patterns, which we can do very economically.

V5: Have you ever taken your furniture outside the context of a project and began to market it as a stand-alone product?

LS: No, we haven't, but right now we are working with Dave Scott, who we do the steel work with, we are just starting now to do some products. We are going to develop a new product line and expand it beyond furniture.

V5: It seems that you hire lots of young people into

答：是的。现在还在施工。旧金山的丹·索罗门（Dan Solomon）胜出。但是我们相信我们的策略是正确的——让客户拥有自己的房屋而不是租赁。我们给每个单元房屋配备了私人花园而不是公共阳台，不过这增加了项目造价，从而比丹·索罗门的设计实施起来困难。我想这也许是我们为什么没有取胜的原因。但是我们依然认为自己的设计更加出色。

问：第一州际银行承担了你们的开发费用吗？

答：我们的确获得了一些钱来准备设计而不是像大多数竞标者得自己掏钱来准备设计。这意味着我们损失得更少了。（笑）

问：我们刚采访了参加北京竞标的Yo Hakomori和他的团队HPST，他们表示时间紧而且花费昂贵。这是一个相当特殊的任务，你的工作室经常参加这类竞标吗？

答：我们常参加。但是是有选择性的。我们会挑选自己真正感兴趣的竞标项目。我想这对工作室是有益的。

我们谈到项目承包者如何成为我们团队的一部分，这通常是在预算上下功夫的结果。我们经常在设计的第一周开始施工，而且团队协作地非常好。"CLICK3×LA"项目始于透明墙的概念，所有的办公室都包上了波状玻璃纤维，我们希望它能够发光。当走进大的电影公司时，你会发现他们的办公室阴暗无光，这是一个很糟糕的工作环境。办公室是非常重要的工作地点，自然光线的进入是这个环境的一部分。我们要寻找一种设计方法来保护隐私，这对他们非常重要，同时保证足够的光线。墙壁就像一个透明的滤光器，当光线进来时，整个空间就亮了。

问：你们也设计地毯吗？

答：是的。我们设计地毯、咖啡桌和接待桌。我们把普通的地毯拼接起来，使它图案化。这样做非常节省开支。

问：你们是否曾经把自己设计的家具作为独立于项目的产品推到市场上去？

答：没有。我们还没有这样做。不过我们现在正在和戴夫·斯科特（Dave Scott）一起进行钢制品的设计开发。我们刚开始做一些产品设计。我们打算开发一条新的生产线并扩展家具以外的领域。

问：你的工作室任用了很多年轻人，是吧？

答：是的，我们大都是年轻人。其中一

your studio.

LS: Yes, we are mostly young people. Some of the people I have known for a long time. For instance, Heather and Jackson were students of mine, and Tim is an SCI-Arc graduate and the most experienced person. Joe is an engineer and fresh out of school.

V5: What advice would you give a student?

LS: I tend to look for people who can acquire experience. There are some people who get ten years of experience in ten years; I look for the people who can get ten years of experience in two years. So being a thinker is important, I don't necessarily look for proven skills or experience, but people who think critically about what they do. I want people in our office who contribute and not just do whatever we tell them to do. To be able to bring to the table ideas and challenge our ideas and test them. We are also small enough that we care about architecture and try to operate it like a studio. You can see in our studio there is not a single door. (laughs) So our private meeting space is outside in the parking lot, that's where we conduct private meetings. (laughs) Everyone here can draw and we still draw a lot even though everyone has a computer. Freehand drawing is still an integral part of what we do. Everyone here is very versatile. We operate like a little kid's soccer team! The whole team follows the ball, so that's how we go, everyone is on a project then we jump to the next project. It is partly because we are in constant charted mode because of how fast we have to do projects. It is very intense, almost like doing a studio project, so we need people who can think quick and develop ideas quickly and implement them. I wish we had projects that we could spend two years on, but most of them are very short and intense, but extremely gratifying. We will work on a project and in three months or less, see it completely built. There is a certain beauty to that as opposed to a big house that takes five years to finish. One of my first collaborations was with David Hertz in 1989. He fabricated a concrete stair for us. This was my first project in Los Angeles that I was really excited about. This was the typical student project in the sense that I got so into it and detailed everything. What happened in the course of this project was the owner; a film director and a

些人我已经认识了很久了。例如：海德（Heater）和杰克逊（Jackson）都是我的学生，蒂姆（Tim）是南加利福尼亚建筑学院的研究生，他的经验最丰富。乔（Joe）是一名工程师，刚从学校毕业。

问：你对学生们有什么建议吗？
答：我在寻找有经验的人，有些人在10年间获得10年的经验；然而我需要找到在两年中就能得到10年经验的人。所以，善于思考非常重要。我不一定需要那些已被证实的经验或者技巧，我需要的是能够批判地看问题的人。我希望我的员工能够做出独特的贡献，而不仅仅去完成任务；他应该能够提出想法、挑战我们的想法并且去验证它们。同时，我们的工作室很小，以至于我们只关注于建筑本身，并努力按照工作室的模式去运作，你可以看到我们的工作室连一个门都没有（笑）。所以我们的私人会议在外面的停车场里进行（笑）。这里每个人都使用手绘，即使大家都有电脑。徒手绘制是我们工作中不可缺少的一部分。这里每个人都多才多艺。我们就像是一个小孩子的足球队！整个队伍跟着球飞跑。每个人都全力以赴地投身到项目中去，然后再跳到下个项目。因为我们需要极快地完成项目，所以大家都像永动机一样工作。工作非常密集，所以我们需要那些思维敏捷、能够快速地发展方案并有实施能力的人才。我当然希望我们能用两年充分的时间来完成项目，但是大多数的项目周期都非常紧而且要求高。我们得在3个月甚至更少的时间内完成一个项目。比起用5年完成的大房子，它具有某种美感。我第一次与人合作是在1989年，合作对象是戴维·赫兹（David Hertz），他帮我们制作一个水泥台阶。这是我在洛杉矶的第一个项目，它让我感到兴奋。这是一个典型的类似学生项目的工作，我为此付出了巨大精力并且事必躬亲。然而在项目运作期间，我们的客户——一个非常有权势的电影导演——不断地为难我们。他扔掉了我设计的很多东西，破坏了我原先构想的形式，这使我非常生气。然而到了最后，项目居然完成得更

very high-powered guy kept stripping it. I was very angry that he was destroying my building by taking out all the stuff that I had designed. But in the end, it was a much better project, because he reduced it to it's essence and I was surprised. It was a great lesson for me because I could then see the beauty in simplicity.

V5: You work with metal fabricators and all different types of craftspeople but David Hertz is an architect. Do you often work with other architects?

LS: I have always loved David's product. Then when I came to Los Angeles I was thinking of the possibilities of Syndecrete and was able to develop them in this project. We had to go very quickly. The client wanted a really strong image, so we took a simple idea of the stair to connect the two levels. I though it should be a very powerful mass, yet how we could do this in budget and on time. I approached David with the idea of a spiral stair and said I need to make this thing while the building is under construction, so when the building is done, we can just assemble it there. We came up with the idea of three pieces, a sort of radius rail piece, and a tread and tread support. David worked with us and we made one prototype and then adjusted it. David went to town and had his guys start fabricating piece after piece, so there were about twenty sets of components. Then when we were ready they brought the parts to the job site and we bolted them into place. That was the first project we collaborated on.

V5: The heavy base of concrete stair and the very light system of metal for the walk is a very nice contrast.

LS: That is basically the idea. The concrete stair ties two levels together so they can interface. It really is an exciting place when they are doing casting calls because there is this incredible activity. I think what I am most proud of is that in almost ten years they have not altered this project in anyway. I'm most pleased that most of our projects have remained unaltered. It says that people like their environment.

V5: When you went back is there now a patina to the materials?

LS: Yes, in some ways it feels better because the Syndecrete material is wearing at the stairs. It is aging nicely now. It feels like an old place now. (laughs)

V5: You have taught at a range of different schools here. What do you think of the schools in the Los Angeles area? How much time do you spend teaching?

LS: I've taught at SCI-Arc, University of Florida, Mississippi State University, University of New Mexico, Otis College, and Woodbury University. At Woodbury I did a joint project with Jennifer Siegal, where we used a sixty-seven foot trailer that was donated by the Salvation Army for a client of mine, a non-profit group. I approached Jennifer and we

好，因为剩下的东西都是精华，这使我非常惊喜。这的确是个教训，我从此看到了简约中蕴涵的美。

问：你和金属制造商以及各种工艺师一起工作，但是戴维·赫兹是一个建筑师，你经常和其他建筑师合作吗？
答：我一直很欣赏戴维的产品设计。当我来到洛杉矶的时候，我正在思考采用合成可循环材料的可能性，并打算将其发展并运用于项目。我们的动作必须要快。依据客户对形体坚固性的要求，我有了一个关于连接上下两层的台阶形式的大致想法，我想它应该是一个坚固的整体，但如何依照时间和预算来实现它是个问题。于是，我和戴维讨论螺旋式楼梯的可行性并且告诉他：我需要在项目施工阶段提前做好它。当建筑一竣工，我们就可以将其组装好。我们最终的方案是楼梯有三个组成部分：辐射型栏杆、台阶和台阶支撑。我们一起做了个模型并修改到理想的形式。戴维带领他的手下回到城里逐一制作楼梯的各个部分。整个楼梯分成了２０多个部分。完成后我们把各个部分带到施工现场并安装成功。这是我们第一次完满合作的项目。

问：厚重的水泥台阶踏面和轻巧的金属扶手真是形成了鲜明的对比。
答：这正是我们的理念所在。水泥台阶把两层连接在了一起。当他们在这里进行电影演员选角活动的时候，场景真是妙极了。使我感到骄傲的是，他们10年都没有改变建筑的原貌。我对此感到非常高兴。听说人们都很喜欢那里的环境。

问：当你再次回到那里的时候，有没有看到材料生锈呢？
答：是的，在某种意义上，它更美了。楼梯上的合成材料虽然有些磨损，但是却越来越漂亮，看上去更像个老地方。（笑）

问：你曾经在不同的学校任教。你怎么看洛杉矶的学校？在教学上你大概花了多少时间？
答：我曾经在南加利福尼亚建筑学院、佛罗里达大学、密西西比州立大学、新墨西哥大学、奥蒂斯艺术与设计学院和伍德伯里大学执教。在伍德伯里，我和珍尼弗·西格尔（Jennifer Siegal）合作了一个教学项目。我们得到了一个67英尺长、由非赢利救世军组织捐赠的活动拖车，我和珍尼弗用一个夏天把它改造成了一个教学工作室建筑。我对教学很有兴趣，以上只是其中一种方式。

talked about doing this as a studio and we did in one summer! So I'm interested in giving back as well and this is sort of one form of it.

V5: I think that is an amazing project, I have walked through some of the Los Angeles junior high and high schools and there are all these temporary buildings and really awful trailers set up that the students must spend six hours a day in. There is no quality grade given to schools.

LS: In the case of that trailer, for thirty-five hundred dollars, you can make these incredible things out of a very simple project with a singular move. So things become incredibly rich with very little in terms of the form. I think of the Salk Institute, if you look at the drawings for that and you give those to a student today, and present this as one of your second year projects, you would be failed as a student. But go see the building, it is awe-inspiring. There is a difference between looking at the form and experiencing the place.

V5: I have always thought that with Baragan's work. When you look at his work in plan, it does not seem extraordinary; there are not the clues that say this is going to be an amazing piece of architecture. Yet when you see it there is something that is just breathtaking. It is a little scary because we have to make so many judgments from plan form and sectional form and models and it is as though someone is seeing things that we are not able to.

LS: It's hard especially if you are early in your career. Our plans of our projects tend to be pretty boring, the plan is generally orthogonal. I guess the old adage of the plan as the determinate of form for us is really not of significance or an issue. We often get times when our clients will say, can't you rotate this nor do something in plan, and it's always work to convince them not to worry about what the plan is. That's not really important. It's the space that is really quite important, and the plan is just an outcome of your intentions. I try to teach one class a year, although I have taught the last three semesters in a row, which is a bit too much for me.

I like the energy of being around students and students contribute to make your work better. One of the beautiful things about Los Angeles is there are a hand full of really good schools and there is the opportunity to teach in several places.

V5: Do you find your interest more in line with certain programs.

LS: No, not particularly, since my interests vary so much. I am interested in the idea at the moment and although there are many common threads in our work, my interest lies in a wider array of experiences. What I do like tremendously about Woodbury University, the students there are incredible in the fact that they are extremely hard working. Most of the students are there because of circumstance, and not because of a great desire to

问：我想这真是一个了不起的项目。我曾经在洛杉矶的中学里看到过临时建筑和糟糕透顶的活动房，学生们一天得有6个小时呆在里面。这里没有好的建筑类型提供给学校。

答：这个只花费了3 500美元建成的活动房项目让我们从一个非常简单、奇妙的项目里得到了令人难以置信的东西。所以说形式语言的简化使事物变得更加富于内涵。如果你在今天看到索尔克学院项目的草图并把它交给学生，让他将草图作为二年级课程设计的作业交给老师，那么你恐怕会失败的。但是，你让学生去看这个建筑，它却是令人惊叹不已的。从图纸上看一个建筑和到现场实地体验之间有着很大的不同。

问：在思考这个问题的时候，我总是在想巴拉甘（Baragan）的作品。他的作品平面图看上去并没有什么特别之处，并没有线索显示这将会是一个令人惊喜的作品。但是当你亲眼看到它，它却让你感到激动不已。这有点令人头疼，因为我们必须从平面图、剖面图和模型中作出很多的判断，就好像别人看到了我们无法看到的东西。

答：这的确很困难，特别是在你从事这个职业的早期。我们的项目平面图就很乏味，通常都描绘得很简单。我觉得图纸并不一定很重要。客户经常说："你们不能多画点设计图吗？"但我们常劝他们不用担心设计图的问题，因为那并不是真正重要的事。空间本身才是最重要的，设计图只是你意图的结论。我每年争取教一门课，我刚连续教了三个学期，这对我来说太重要了。

和学生们在一起使我充满了活力，和学生的互动也使我的工作更出色。很美妙的是，洛杉矶拥有很多很棒的学校，这为我提供了在不同的地方任教的良机。

问：你的个人兴趣集中在某类项目吗？

答：不，我的兴趣很广泛。我对瞬间的创意充满兴趣，虽然我的作品中有很多普通的思路，但是我的兴趣在于更广泛的体验。我特别喜爱伍德伯里大学，那里的学生非常努力。大多数的学生选择伍德伯里是因为那里的氛围。他们如饥似渴地工作让我十分地感动。这使我想起：在过去，学生们趋之若鹜地进入南加利福尼亚建筑学院学习，因为那里有出色的教学力量，学生们动力十足。如今，伍德伯里大学则是出类拔萃的领头者。

问：听上去有点当仁不让么。（笑）

答：学校非常进取。聪明的孩子们非常热衷于建筑事业。

go to Woodbury. They are hungry and hardworking and remind me in some ways of SCI-Arc in the old days where people went there because they could get into the school. There were incredible faculty and the students were driven, and I think Woodbury is that school now.

V5: It is scrappy. (laughs)

LS: Very, very scrappy, blue collar. Bright kids and very hungry for architecture.

V5: You get the sense teaching there that for about half of the students, it is their family's first generation into higher education. They are very hard working and a little scared.

LS: Right. Maybe that's partly why I like it there too because I'm the first generation, college educated in my family. Perhaps there is some kind of relationship.

V5: Where did you go to school?

LS: I went to school at the University of Florida. My father is an Italian immigrant and most of my family barely finished high school. When I grew up, we really didn't have too many books around the house other than Sports Illustrated. I always knew from a young age that I wanted to become an architect, but mainly because my father, who worked for the post office, used to do small construction jobs such as additions. I would hang out with him and get in his way, and that's what I thought architects did. (laughs)

V5: Maybe it is what architects do!

LS: I always wanted to become an architect and the interesting thing was my father had also always wanted to become an architect, but because of having four children, it just never happened.

V5: Was your family supportive of your education?

LS: Yes, but frankly, my parents really didn't know what to do. They were supportive about going to college, but had no idea what you do to get accepted to college. I wasn't the most stellar high school student, although I knew what I wanted to do at an early age. So I went to the local college my first year and took the only class that the college offered that resembled architecture, I think it was a drafting class. I was lucky enough that the guy who taught it asked me what I was doing there. I said I was going to be an architect, to which he replied, "You are in the wrong place". (laughs) He was the one who told me what I needed to do in order to go to architecture school and educated me on the process of becoming an architect. I didn't really wake up until my first year of college. I wound up transferring very quickly to Florida and upon graduation, I moved to New York and worked for Paul Rudolph. He was the guy who was a very big influence on me while I was in school.

V5: That's quite of jump. Did someone help you in college to get into the Rudolph office?

LS: No, but I worked summers with an architect, who I sort of forced myself upon until he

问：那里有一半的孩子都是你的学生。孩子们大都是他们家庭中第一代获得高等教育机会的人。他们努力学习，甚至很好胜。
答：对。我喜欢那里的部分原因也在于我是我家第一代受过高等教育的人。可能这有点关系。

问：你在哪儿上的学？
答：我上的是佛罗里达大学。我的父亲是意大利移民，我家里大部分的人都没上过高中。我小时侯，家里除了体育画报就没什么可以读的书了。我父亲在邮局工作过，他曾经从事过小型的建筑工作。在他的影响下，我从小就打算从事这个行业。（笑）

问：也许这就是建筑师所做的事。
答：我一直想成为一名建筑师。有趣的是：我父亲以前也希望做建筑师，但是因为有4个孩子，所以没有实现梦想。

问：是你的家庭承担了你的教育费用吗？
答：是的。但是坦率地说，我的父母能够支持我上学，但是对我能否被学校录取却没有数。虽然我知道自己的目标是什么，但是在中学的时候我并不是最出色的学生。所以，我在大学一年级上了本地的地方学院，我参加了学院惟一和建筑有关的课程——工艺课。幸运的是，任课老师问我为什么来上课，我说我打算成为一名建筑师，他回答说："你来错地方了。"（笑）他建议我去上建筑学院，并且告诉我成为建筑师的过程是怎样的。我好像直到大学一年级才明白过来。于是我很快转学到佛罗里达大学并且在那里毕业。后来我来到纽约为保罗·鲁道夫工作。他对我的影响很大。

问：这的确是个飞跃。你所在的学校是否有人帮助过你进入鲁道夫事务所工作？
答：没有。但是我在暑期为一名建筑师工作过，他对我后来的工作起到了积极作用。

问：你在他门前的台阶上睡觉了吗？
答：我在高中打棒球的时候，球场挨着一个住宅小区。我们的球越过了栏杆，我得去把球捡回来。当我看到小区里奇妙的建筑时，简直被它迷住了。那是用预应力混凝土制成的。后来我查到了这个建筑的设计者。我说:我要为这个人工作。他最后也任用了我。他的名字是吉恩·李迪。我整个暑假都为他工作，直到暑期结束。直到今天我们仍是很好的朋友，他正好是保罗·鲁道夫在佛罗里达的第一个雇员。于是，吉恩向保罗推荐了我。

acquiesced.

V5: Did you sleep on his doorstep?

LS: When I played high school baseball, our playing field was up against a residential area, and balls were hit over the fence, which I would have to go get. I remember seeing these incredible houses and they were made out of pre-stressed concrete and I used to look at them and go, "Wow". I then found out who the architect was and I said this is the guy I'm going to work for. He wound up hiring me; his name was Gene Leedy. I ended up working for him in the summers and to this day we are still very good friends. He was Paul Rudolph's first employee in Florida, so there was that connection there as well and he put in a good word for me.

V5: What was that like being in Paul Rudolph's office?

LS: I pulled more all nighters in the two years that I was there than in my four years in school. It was hard work, but extremely gratifying. There were certain things in his office that I learned such as how to develop a method of design. One of the great experiences was a house in Jacksonville, Florida that he did called the Biome House, incredible concrete block house, and I always loved that house. So one of the first things I did when I started work there was I went to the basement in the shop in SoHo and found the drawings for the house. I pulled out all the drawings and when I opened them up, I was in complete shock. He kept very good records, so when I looked at the very first sketches of the house and it was the most god-awful design I had ever seen. (laughs) But then when you look through it, you can see the evolution and at that point I realized there might be hope for me. I had thought his talent was god given and, with a stroke of genius had just come with the design. It was a lot of hard work and evolution; at that point I said there might be a chance for me. At that point I became convinced that if you worked hard, you could better yourself significantly.

V5: There is always a kind of honesty and presence about Paul Rudolph's work.

LS: That was the way his lifestyle was and the way his office worked.

V5: Did everyone work together?

LS: His office was a big studio, when you got off the elevator you were in the middle of the office. I had always been under the impression before I went to work there, that it would be a corporate place like I.M. Pei's studio or SOM. It was completely opposite, it was like going into a SCI-Arc studio, and it looked just like that. You got off the elevator on Fifty-seventh Street; you opened the elevator, and were in the middle of a bunch of desks. There was nothing pretentious about it; it was a studio, not really an office.

V5: Where there hierarchies within the studio?

LS: No, there were only about eight people there

问：你在保罗·鲁道夫那儿的工作状况怎样？

答：我在那里工作的两年间开夜车的时间甚至超过了上大学的四年。工作真是辛苦，但是非常令人满足。我从那里学到了很多东西，比如如何开发设计方法。其中令人难忘的工程是在佛罗里达的杰克逊维尔（Jacksonville）的"生物群住宅"（Biome House），那是出色的混凝土住宅，我一直都非常喜欢那个房子。我上班做的第一件事情就是到位于苏荷的工作室的地下室找到房屋的设计图。当我打开所有的图纸时，整个人都惊呆了。记录保留得很完整，看到第一张草图时我觉得它简直就是神来之笔。（笑）当你仔细地整个看完时，你会发现革新的路径，这一点让我领悟到了我的希望所在。我曾认为他的天赋是神给予的，这是一个天才的设计。当你看到这大量辛苦的工作和革新时，我看到了我的机会，这让我坚信：如果你努力工作，你就可以让自己变得更好。

问：保罗·鲁道夫的作品总是带着一种诚恳和风度。

答：他的生活态度和公司都是这样。

问：所有人都在一起工作吗？

答：他的工作室很大，从电梯口出来就到了办公室中央。我原来认为他的办公室应该像贝聿铭或者SOM的工作室，但是事实上是完全不同的，它更像是南加利福尼亚建筑学院的工作室。当你走下57大道的电梯，打开电梯门，站在书桌前，那种感觉太像了。它一点也不自命不凡；它就是个工作室，不是真正的办公室。

问：工作室的等级制度如何？

答：没有等级制度，只有8个人。我只是其中一个，工作时间最长的一位在那里工作了大概20年，另外有一位工作了10年。真的没有什么等级划分。他给予每个人足够的体验和机会。当然，他是一个严厉的人，不过幸运的是，我和他相处得非常好。

问：他是个很安静的人吗？当我还是学生的时候，在一本《与建筑师对话》的书中惟一一次见到他的样子。

答：是的，他非常安静，但是有时候也很活泼。

问：他是如何应对建筑潮流的改变的呢？比如从"现代主义"向"后现代主义"的变化？

while I was there and one guy had been there for about twenty years, another person ten years. There was really no hierarchy, he gave everyone as much experience and opportunity as you would take. He was a tough guy to work for but fortunately I got along with him very well.

V5: Was he fairly quiet? The only personal view I have of him is from the book "Conversations with Architects" that I read when I was a student.

LS: Yes, he was very quiet, but he could also be very volatile.

V5: How did he deal with the changes that happened in architecture, as styles and preferences moved away from the modern idiom and into "Postmodern"?

LS: He stayed with what he believed in. In fact, the AIA tried to give him the gold medal and he refused to join the AIA. Even the people who had worked for him, past employees and so forth, many who were very close to him, especially during the early years of his career, said, "We will do all the paperwork, all you have to do is agree to sign up." He wouldn't do it. In some ways he really practiced as an outsider and stuck to what he believed and that was it.

V5: Is it personally difficult for him to see major commissions go to other architects?

LS: No, he was busier later in his life than he was early in his life. He had tremendous projects overseas in Southeast Asia, spending a couple of months there each year. At least to me it never was an issue for him. He never seemed frustrated about recognition. I had been to all his projects from the 1940s, ones he did with Ralph Twitchel, the Cocoon House, which was almost destroyed and I would show him pictures. He was heartbroken; it would crush him to see them in such disrepair. It really bothered him because he truly bled architecture. He was a very interesting guy. After I worked for Paul, I went to graduate school at the University of Florida. Then moved to Vicenza, Italy and was there for about two years teaching and doing research on Carlo Scarpa.

V5: Is there a family relationship there?

LS: No, the name Scarpa comes from a small fishing town just south of Venice, and if you look in the phone book there, Scarpa is the equivalent of Rodriquez in Miami. (laughs) It's not an uncommon name, but I actually did my graduate thesis project on Carlo Scarpa. It happened strictly out of coincidence. The interesting thing was that most everything that was written about Scarpa at the time, even in Italian, was all bits and pieces, nothing that talked about his body of work, and that is what I focused on for my thesis.

V5: There is a strong sense of material quality and detail development that is supportive of the design ideas in both your work. Thank you Larry.

答：他坚信自己的追求和判断。事实上，美国建筑师学会试图授予他金质奖章，然而他拒绝参加美国建筑师学会。曾经为他工作过的、和他关系亲密的员工，特别是他职业生涯早期的雇员曾对他说"让我们来做所有的案头工作吧，你需要做的就是签个字。"但他拒绝这样。有时候，他就像一个门外汉一样去尝试、实践很多事情，然后努力完成他所坚信的一切，就是这样。

问：对他来说，看到大的项目被其他的建筑师获得，会很难接受吗？
答：不会。他比年轻的时候更忙了。他在东南亚有大量的项目，每年要花费好几个月去完成。至少在我看来这不是个问题。他从来没有因为是否被认可而沮丧过。我寻访过他在20世纪40年代后的所有项目。他和拉尔夫·特威切尔（Ralph Twitchel）合作的科库恩住宅（Cocoon House）基本上被破坏了。我给他看图片时，他非常伤心，这是惟一能击垮他的事情，因为他把心都交给了建筑。他真是个非常有趣的人。在我为保罗工作之后，我考上了佛罗里达大学的研究院。然后去了意大利维琴察任教并从事对卡尔罗·斯卡帕（Carlo Scarpa）的研究。

问：你和那里有家族渊源吗？
答：没有。斯卡帕的名字来源于威尼斯南部的一个镇。如果你查查电话黄页会发现斯卡帕像迈阿密的罗德里格斯一样普通。（笑）这个名字太常见了，不过我的硕士论文的确是有关卡尔罗·斯卡帕的。这就是个巧合而已。有意思的是，有关卡尔罗·斯卡帕的作品即使在意大利也没有人谈及。而我的论文则着重研究他的情况。

问：我认为高质量的材料和细节开发都是支持你们两位建筑师设计理念的重要元素。感谢你接受我们的采访，劳伦斯。

Below: model of the Dwell project
Dwell 项目模型

Contents

Introduction with Article Review, Page 7.
Project Survey **Colorado Court** Page 23. COop Editorial Page 41. Nascent Terrain Page 59. Diva Page 67. Orange Grove Page 77. North Point Page 85. Jigsaw Page 95. Solar Umbrella Page 109. SM College Student Service Center Page 117. Bergamot Loft Page 125. Vail Grant House Page 139. The Firm Page 149. XAP Page 159. Reactor Films Page 171. Chronology Page 181 & Bibliography

Colorado Court

科罗拉多法院公寓
圣莫尼卡，加利福尼亚州，2002

科罗拉多法院公寓是美国首座能源自给的建筑。它和大多数传统项目不同的地方在于：它结合了超越标准方法的能源节省策略，优化建筑性能并且保证在整个施工阶段减少能源的使用。科罗拉多法院公寓的规划和设计来源于对太阳能设计策略的充分考虑。这些策略包括：考虑控制太阳能冷却负荷的情况下为建筑选址并定位；考虑建筑暴露在全方位风吹情况下的定型和定位；把建筑塑造成可以引入自然通风的形式；设计能够将光照最大化的窗户；设计南面遮光窗户，减少西窗玻璃反光；设计能最大限度保持通风的窗户；加强光照和自然空气流动的内部规划。

科罗拉多法院公寓含有多种艺术和工艺技术的特色，是可持续性能源供应和利用的范例。这些技术包括用于满足大楼基本电力负荷和热水供应的天然气涡轮引擎恢复系统，以及一个能够提供高峰电力负荷的太阳能电子板系统，这个系统安装在建筑的表面和顶端。这个共同发电系统将天然气转换为电力，来满足建筑基本负荷需要，并且能够利用剩余热能全年供应热水并在冬季供暖。相对于原有设施只能节省30%以下的能源，新的天然气系统能够节能85%以上。太阳能光电系统能够产生绿色电力，不会给周围带来任何污染。太阳能光电板是建筑外壳的一部分，没有被使用的电在白天被送往输电网，到了夜晚再被输送回来。上述系统的成本只要不到10年就可收回，每年节省的电力和天然气估计价值超过6 000美元。

建筑材料包括：
大多数材料更优于圣莫尼卡市对可持续性建筑的现

Colorado Court

Location of Project: 502 Colorado Avenue, Santa Monica, California, USA
Client/Owner: Community Corporation of Santa Monica
Total Square Footage: 30,150 square feet

Awards: 2003 National AIA Design Award, 2003 AIACC Award, 2003 AIA/LA Award, 2003 Rudy Bruner Prize, 2003 World Habitat Award Finalist, 2003 AIA COTE "Top Ten Green Building" Award, SCANPH "Project of the Year", 2002 Westside Urban Prize, 2003 AIA PIA National Housing Award

Project Team: Lawrence Scarpa, AIA - Principal- in-Charge, Angela Brooks, AIA - Project Architect, Gwynne Pugh, AIA, Anne Marie Burke, Heather Duncan, Vanessa Hardy, Tim Peterson, Ching Luk, Jackson Butler, Steve Kodama, FAIA

Project Energy Engineer: Dr. John G. Ingersoll of Helios International Inc.-- Energy Efficiency Measures, and Distributed Power Generation - Solar PV Power and Co-Generation - for Sustainable Development in the Built Environment.

Structural Engineering: Youssef Associates
Mechanical Electrical Plumbing Engineering: Storms and Lowe

Program:
The program for this single resident occupancy housing project includes:
44 single resident occupancy units (375 square feet max per unit)
Community Room, Mail Room, Laundry Room, Outdoor common courtyard spaces @ ground level and 2nd level
On-grade covered parking for 20 cars
Bike Storage

Colorado Court is one of the first buildings of its type in the United States that is 100% energy independent. Colorado Court distinguishes itself from most conventionally developed projects in that it incorporates energy efficient measures that exceed standard practice, optimize building performance, and ensure reduced energy use during all phases of construction and occupancy. The planning and design of Colorado Court emerged from close consideration and employment of passive solar design strategies. These strategies include: locating and orienting the building to control solar cooling loads; shaping and orienting the building for exposure to prevailing winds; shaping the building to induce buoyancy for natural ventilation; designing windows to maximize daylighting; shading south facing windows and minimizing west-facing glazing; designing windows to maximize natural ventilation; shaping and planning the interior to enhance daylight and natural air flow distribution.

Colorado Court features several state

Below: exterior view of the Colorado court

科罗拉多法院公寓建筑外观

of the art technologies that distinguish it as a model demonstration building of sustainable energy supply and utilization. These technologies include a natural gas powered turbine/heat recovery system that generate the base electrical load and hot water demands for the building and a solar electric panel system integrated into the facade and roof of the building that supply most of the peak load electricity demand. The co-generation system converts utility natural gas to electricity to meet the base load power needs of the building and captures waste heat to produce hot water for the building throughout the year as well as space heating needs in the winter. This system has a conversion efficiency of natural gas in excess of 85% compared to a less than 30% conversion efficiency of primary energy delivered by the utility grid at the building site. The solar photovoltaic system produces green electricity at the building site that releases no pollutants to the environment. The panels are integral to the building envelope and unused solar electricity is delivered to the grid during the daytime and retrieved from the grid at night as needed. These systems will pay for themselves in less than ten years and annual savings in electricity and natural gas bills are estimated to be in excess of $6000.

Construction Materials include:

Most materials exceed the City of Santa Monica Standards for Sustainable building and comply with the U.S. Green Council Building Standards
Concrete Masonry Unit: structural walls, ground level
Building Wall Insulation System: Walls meet R23 value- plaster finish with blown-in recycled insulation.
Roofing/Insulation System: High density foam insulation and High performance SBS modified bitumen membrane roofing and RSD insulation
State-of-the-art solar photovoltaic integrated wall panel system
Glazing High Efficiency Dual Glazing ("low-e") with krypton gas
Exterior Finishes: Concrete Masonry Unit (CMU) face block, integrally colored cement plaster, galvanized sheet metal

Special thanks:
City Of Santa Monica - Construction And Permanent Funding
Santa Monica Redevelopment Agency
Southern California Edison - Funding Assistance And Environmental Systems
Robert C. Johnson - California Energy Coalition
Craig Perkins City Of Santa Monica Public Works Department
City Of Santa Monica Environmental Dept

Colorado Court

有标准并严格遵守美国绿色建筑标准。

混凝土：结构墙、地基
建筑墙壁隔热系统：达到R23值的带有通风循环隔热性能的石膏面墙壁
屋顶/隔热系统：高密度泡沫隔热和高性能SBS改性沥青涂膜层式屋顶，RSD隔热材料
太阳能光电结合墙板系统
高效防辐射双层玻璃
低辐射节能玻璃和氪气
外墙材料：混凝土面砖、混合水泥石膏、电镀金属板

Above: rendering of the Colorado Court and the porch view of the completed building

科罗拉多法院公寓效果图和外廊景观

Interview with Lawrence Scarpa

Q: How did you get into affordable housing?
A: We always wanted to do it and one of our partners, Angela Brooks, had won a PA award for her housing work. We tried very hard for many years but it was a difficult market to get into. So in 1996 we formed a partnership with a San Francisco architect, Steve Kodama, who has 35 years experience doing housing. Pugh-Scarpa-Kodama, is a separate firm that focuses on affordable housing. Kodama grew up in LA and he wanted to be more active here. We had a mutual friend who helps cities put together housing programs and he introduced us. I think we've done about 10 projects together.

Q: Do you worry about diluting or confusing the brand of your firm?
A: People in the affordable housing sector don't care about the brand thing.

Q: What was your motivation?
A: We believe in giving something back, do something for the greater good. We have a lot of film industry clientele. So, it's a way to look at the other side of architecture, but I think they turn out to be one in the same. You can bring the same ideas to affordable design as to offices for movie stars. I don't think tight budgets preclude good design.

Q: How did you get so much sustainability into Colorado Courts, an affordable housing project?
A: We have always been interested in sustainability. Six or seven years ago we designed with Tony Loui the only totally solar powered electrical vehicle charging station in the U.S. next to Santa Monica City Hall. To do that we had to be pretty creative about our strategy for funding. We found public money nobody knew about.

For Colorado Court, Santa Monica provided the land on a long term lease and provided construction financing with the stipulation that the project would be green. However, there were no guidelines as to what that meant. We could do what we wanted as long as it didn't cost any more. Of course it did, but we had to find the resources. Our strategy was two pronged.

In affordable housing there is no real resource for additional funds. In this case they set aside a larger than normal project contingency. Our strategy was to design a super simple building, so simple that we would minimize potential change orders that would eat into the contingency. Every unit stacks vertically. Variety is in the horizontal elements. We used the balance of the contingency for green items at the tail-end like natural linoleum, formaldehyde free cabinets, paints with low VOC, recycled carpet, and high efficiency refrigerators.

The other strategy was to look for money like we did at the electric vehicle charging station. We

问：你们如何进入经济型房屋建设的市场？
答：这是我们一直希望能够做到的。我们的一个合作伙伴安吉拉·布鲁克斯曾经因该类建筑作品而获得PA建筑设计奖。我们为这个很难进入的市场努力了很多年。1996年，我们与旧金山的建筑师史蒂夫·卡达马（Steve Kodama）建立了合作关系。卡达马在住宅建筑行业有35年的丰富经验。皮尤+斯卡帕+卡达马是一个独立的专注于经济型房屋建设的设计事务所。卡达马希望能够在洛杉矶有更好的发展，我们有一个从事建筑项目交流的共同朋友，是他介绍了我和卡达马认识。现在我们双方已经顺利合作了大约10个项目。

问：你们会担心这会弱化或者混淆你们公司的品牌形象吗？
答：在经济房地区生活的人们不在乎品牌的东西。

问：你们的动机是什么？
答：我们相信在舍掉一些东西后，可以做更好的事情。你可以从我们很多的电影界客户那里看到建筑的另一面，但是我认为展现出来的东西其实是一致的。同一种概念，你可以把它运用于经济型住房，也可以运用于电影明星的办公室设计。我认为有限的预算与好的设计并不矛盾。

问：你们是如何把这么多可持续性因素注入经济型房项目——科罗拉多法院公寓呢？
答：我们一直对可持续设计感兴趣。六七年前，我们和托尼·路易(Tony Loui)设计了美国第一座位于圣莫尼卡市政厅旁边的太阳能电动汽车充电站。为了完成项目，我们使用了非常有创意的策略而获得了资金。我们甚至能够找到根本无人知晓的基金。

在科罗拉多法院公寓项目中，圣莫尼卡市以租赁的形式提供了土地，还提供了建设资金，并提出了此项目的绿色协定。但是对于其中的含义，并没有特别的指示。我们可以在不超支的前提下尽情发挥我们的想像力。当然，最后的确超支了一些，不过我们找到了其他的赞助。我们有两种策略。

在经济型住房项目中没有真正意义上的额外资金。那么，大于正常规模的项目在运作上会有困难。我们的策略是：设计大型简约建筑，简约的建筑形式可以把影响工程的潜在风险降到最低限度。每一个部分都是垂直叠加的，而多样性会在平行元素中产生。我们把节省下来的开支用在项目尾期购置环保产

劳伦斯 · 斯卡帕访谈

got some money from the State through DOE in a buy-down program most people are familiar with. We also found a little known source of funding that is now called the Six Cities Program, which sets resources from utility bills aside for clean air projects.

Q: How did you create an "energy neutral" project?
A: A combination of tools. Solar PV panels, a micro turbine, breezeways, and cross ventilation. The south facades have shade and the north facade has glazing. When we designed the project we went under the assumption that we were going to make it happen. We would subtract the element if we didn't. We were well into the process before we knew it was going to happen.

Q: With the new technology did anything go wrong?
A: The solar panels were by 'solarex,' a small cowyacy that was purchased by 'BP Solar' which we were under construction. They quit producing the panels we wanted so we had to redesign the structure in the middle of construction.

Q: What about some of the design elements. What do they do?
A: They are abstract patterns. It is not a machine. We don't live in machines. Those facades are sculptural provide shading and hide downspouts and gutters. We are interested in place making, a space for people. We are not interested in making a machine that is 100 percent efficient. The key is to make a new architecture, a new paradigm. We made some sacrifices in efficiency for how people use and enjoy the building.

Q: Can this become a model for other affordable housing projects?
A: I think it can. This is the perfect scenario. You've got a building type with long term ownership. That is how these energy strategies pay off. This is the tenant population that needs this savings the most. These are the people who can least afford the utility bills. In low income families utility bills might represent as much as 50% of a family's income.

We have to change the way we think. In our society we think about the least possible amount that a project can cost in capital expenditures on day one. Not what does it cost three, five, ten years from now? These kinds of project are an investment in the future. We cannot afford not to do it. We have wars over oil. We have to look at the long term damage of our resource consumption. Interestingly enough, the strong energy savings have proven to be a marketing edge.

The State of California has changed some of the ways they fund projects because of Colorado Court. They view environmentally friendly projects more favorably with points through tax credits. There is much you can do that costs little or nothing more. For example, using polished concrete floors is less

品，包括天然油毡、无甲醛器具、低挥发性有机化合物（VOC）的涂料、环保地毯和高效节能冰箱。

另一种策略是像电动汽车充电站项目那样寻找资金。我们通过能源部（DOE）从政府获得人们比较熟悉的低利率按揭资助。我们还找到了一个被称为"六城市计划"的项目资助，它是一个为空气净化项目免除水电费用的基金。

问：你们如何创造一个"能源中立"的项目？
答：结合各种手段，包括太阳能电光板、微型燃气轮机、有屋顶的通路和交叉通风系统。南面有遮光板，北面有玻璃窗。当我们开始项目设计时，我们就会设想将它实现。如果不这么做，设计元素就会大打折扣。我们在项目开始之前就把这些要素考虑得很充分。

问：新技术的应用在工程中有出现问题的地方吗？
答：从澳大利亚BP太阳能公司购买的太阳能光电板不大符合我们的要求，为此我们在施工过程中对结构进行了调整。

问：请讲讲设计元素和它们的功能好吗？
答：都是抽象的模式。它不是机器。我们不住在机器里。那些墙面具有雕塑感并能将排水管道隐藏在内。我们感兴趣的是为人们创造空间，而没有兴趣去制造一个百分之一百高效的机器。关键是制造一个新的建筑，一个新的范例。为了节能，我们在人的使用和享受方面作出了一定的牺牲。

问：这可以成为其他经济型住房项目的典范吗？
答：我认为可以。这是一个完美的解决方案。采取能源策略的好处是你能够长久地拥有一个住所。住户需要最大限度地节省能源开支，因为有很多人只能付最少的水电费。在低收入家庭中，水电费可能会占家庭收入的50%。

我们需要改变思维的方式。我们的社会应该考虑一个项目每天最少的花费是多少，而不是3年、5年或者是10年会用多少钱。这些项目是对未来的投资。我们担负不起的事我们就不要去做，我们有石油战争，我们不得不面对能源消耗给我们带来的长期的破坏作用。如今，强大的节省能源策略已经处于市场领先的地位，这是足够令人感兴

expensive because there is no need for carpet. High content fly ash in the concrete is stronger and no more expensive. storm water mitigation is a minimal additional cost. Of course, the orientation of the building helps. A huge thing is to scrutinize the engineering data. My partner is an engineer and an architect and so we questioned the engineers closely. We get them to remove some things. All of that saves money.

Q: What has been the reaction from the occupants?
A: They like it. The head of the housing dept in Santa Monica said, "We have done a lot of housing, this is the first one that everybody likes." I hear that and I get a little bit worried. It must be too soft. They had 3000 people on their waiting list for 44 units. The clients are proud that it is an environmentally sensitive bldg. They were surprised how well the tenants like the building.

Q: What is going to happen to affordable housing?
A: We don't have enough. A few years ago I helped start a non-profit housing development corporation (www.livableplaces.org). I think well intentioned people sometimes lose sight of their vision. Livable Places wants to influence policy. We want people with creativity and commitment to have a chance to design, even if they have not done this kind of work. We were frustrated at how affordable housing was developed and how it looked. I wanted to show that it does matter. Most affordable housing developers don't want to do mixed use because how these projects are funded. We are doing mixed use because it makes sense. So we bring together different funding sources and don't use some of the typical sources that try and tell us how to do things. I think this group has the potential for becoming a new model for the development of affordable housing. It seems to be the only way to shake up some of the funding sources. We did this because it seemed to be the only way to make significant change in how we think about affordable housing. We've received almost $1 million in grants so someone must be listening.

趣的事情。
因为科罗拉多法院公寓的成功，加利福尼亚州政府已经改变了一些项目的资助方式。他们对环保项目采用降低税收的优惠政策。你可以做很多事情来节省花费。比如，使用抛光混凝土地板就便宜很多，因为不需要地毯。混凝土中因为粉煤灰含量高，所以比较坚固，而且也不贵。暴雨排放系统只需要很少的额外开支。建筑的朝向也是一个因素。另一件重要的事情是检查工程数据。我的合作人是一名工程师兼建筑师，我们可以直接和工程师探讨问题。我们会根据他们的建议减少一些内容，以节省开支。

问：来自居民的反馈如何？
答：他们很喜欢。圣莫尼卡市房屋部门的负责人说："我们有很多房屋，但只有这个让所有人都喜欢。"有3 000人排队在等44套房子。我们的客户对它是环保建筑这一点很自豪。而人们如此的热情也令他们吃惊。

问：经济型房屋的发展状况怎样呢？
答：还不能令人满意。几年前我帮助建立了一个非赢利房屋开发机构(网址http://www.livableplaces.org)。我认为很多有着良好用心的人没能看到它的前景。这个机构(Livable Places)希望能够对政策产生影响。我们希望那些有创造力和信仰的人能有机会参与设计，即使他们并不从事这个行业。我们为经济型房屋的开发状况感到失望。我想问题就在于大多数开发商不会去考虑资金来源的综合利用，而我们这样做的原因是使它行之有效。所以我们把不同来源的资金汇集到一起，其中具有代表性的资金被存储起来。我想这个机构有潜力成为经济型房屋开发的新模式。这是惟一能够极大改变我们对经济型房屋看法的方式。要知道，我们已经获得了近百万的资助，人们必须对此刮目相看。

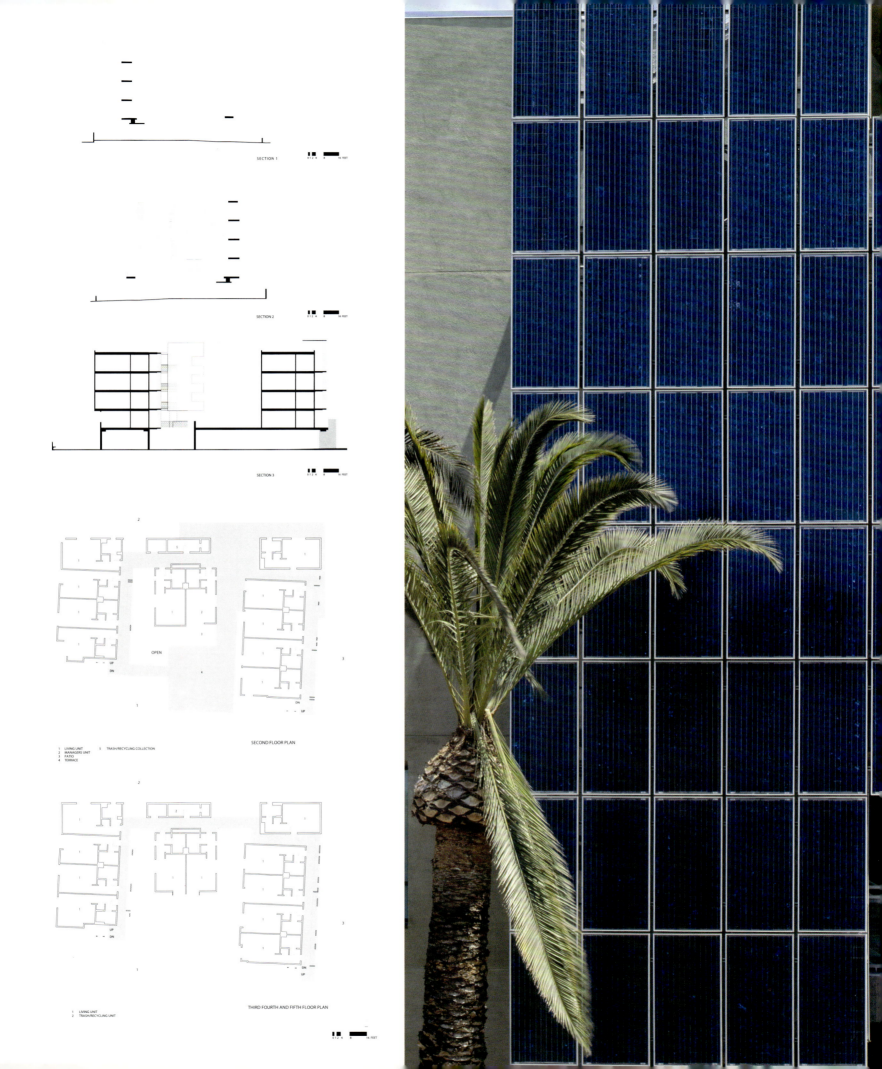

SECTION 1

SECTION 2

SECTION 3

SECOND FLOOR PLAN

1 LIVING UNIT   5 TRASH/RECYCLING COLLECTION
2 MANAGERS UNIT
3 PATIO
4 TERRACE

THIRD FOURTH AND FIFTH FLOOR PLAN

1 LIVING UNIT
2 TRASH/RECYCLING UNIT

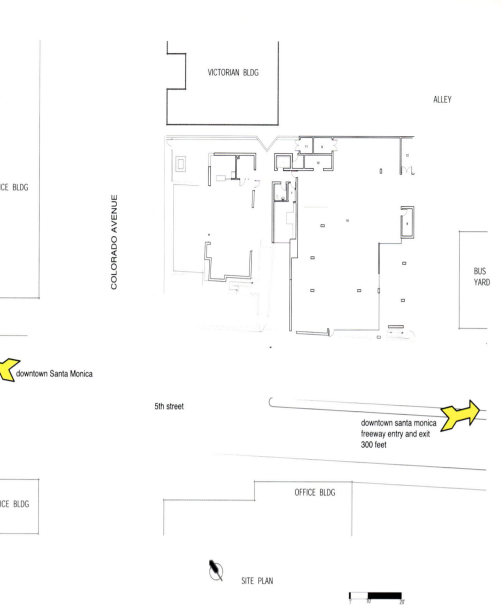

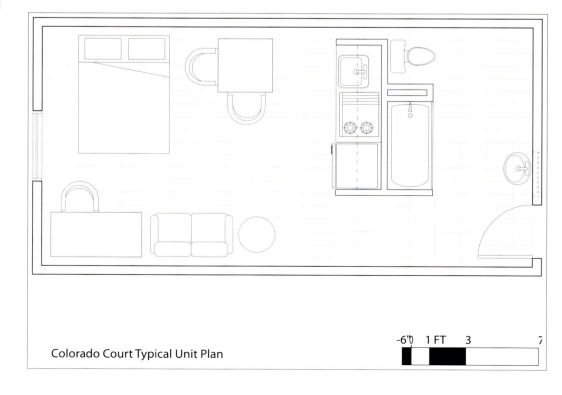

Opposite: renderings of the Colorado Court and ground floor plan

科罗拉多法院公寓效果图和底层平面图

Above and righ: renderings, site map and unit plan

效果图，总平面和单元平面图

Colorado Court Typical Unit Plan

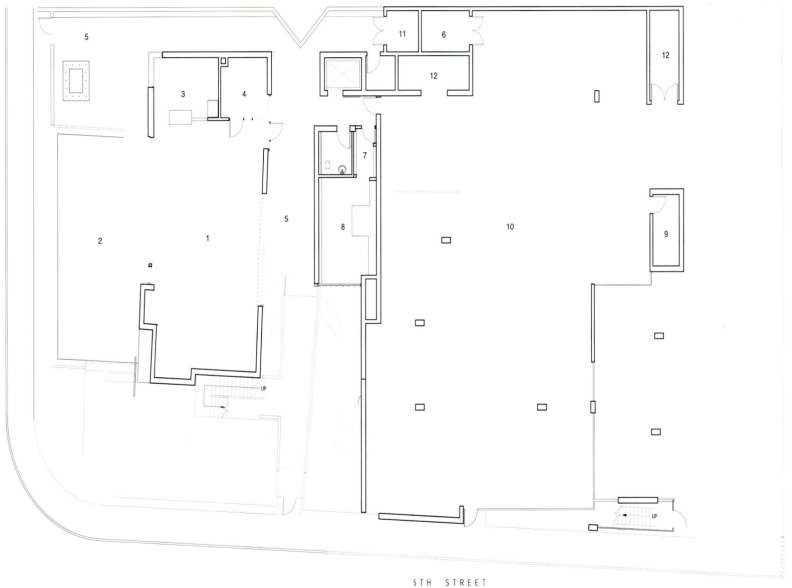

GROUND FLOOR PLAN

1 LOBBY/LOUNGE
2 COMMON COURTYARD
3 KITCHEN
4 OFFICE
5 ENTRY
6 STORAGE ROOM
7 MAIL ROOM
8 LAUNDRY
9 MECHANICAL
10 PARKING
11 ELECTRICAL ROOM
12 TRASH/RECYCLING COLLECTION

Contents

Introduction with Article Review, Page 7
Project Survey Colorado
Court Page 23. **COop Editorial** Page 41. Nascent
Terrain Page 59. Diva Page 67. Orange Grove Page 77.
North Point Page 85. Jigsaw Page 95. Solar
Umbrella Page 109. SM College Student
Service Center Page 117. Bergamot Loft Page 125.
Vail Grant House Page 139. The Firm Page 149.
XAP Page 159. Reactor Films Page 171.
Chronology Page 181 & Bibliography

COoP 剪辑工作室

圣莫尼卡，加利福尼亚州

这个占地4 700平方英尺的项目是对1963年弗兰克·盖里（Frank Gehry）设计的位于圣莫尼卡市中心的一栋商业建筑的改造。这是一个针对原有建筑结构的特别挑战。

项目的客户是芝加哥的一个影视后期制作公司。他们为电视业提供在线和离线编辑、转胶片、特殊视觉和听觉效果制作等各种服务。

像我们的大多数作品一样，这是一个持续进行的对相关材料和技术的研究，以及对现有条件、成规条文和方法的再检验。这引领我们采用一个创新的方式来解决建筑室内设计问题。我们没有预先定义建筑，而是直接凭直觉对空间材质做出反应。

这个工作室的环境设计蕴涵着现实意义的电影或者高速路的体验。设计贴近使用者，提高他们的认知感觉，给他们的体验带来更深的理解和活力。

这个设计检验了材料、形式和体验之间的张力。它的内部空间可以被看作是"呈现在空间表面和包装上内外翻转的建筑结构的交替隐匿和显现"。设计通过新形成的外观层次和塑造把不同的和对立的材料结合在一起。记得著名电影导演阿尔弗莱德·希区柯克充满隐喻的悬念电影吗？在这里一样充满隐喻。空间的虚无和实际外观一样重要，它显现了一种早期的材料使用模式。

项目的设计理念和兴趣点在于超越传统手工艺和提升普通材料的同时，不改变其材料的内在本质。这是着眼于普通事物，并揭示其蕴涵非凡意义的尝

COoP Editorial

Location of Project: Santa Monica, California
Client/Owner: Optimus Corporation
Total Square Footage: 4,700 sq. ft.

Project team: Lawrence Scarpa, AIA - Principal-in-Charge. Peter Borrego, Angela Brooks, AIA, Silke Clemens, Vanessa Hardy, Ching Luk, Tim Petersen, Gwynne Pugh, AIA, Bill Sarnecky, Lawrence Scarpa, Katrin Terstegen.

Engineering: Gordon Polon
Photography: Marvin Rand

The design of this 4,700 square foot tenant improvement evolved from the unique challenge to remodel an early 1963 Frank Gehry designed commercial structure located in the heart of downtown Santa Monica.

The client is a full service post-production facility based in Chicago. They provide every aspect of the TV commercial making process, including creative on and off line editing, film transfer, special, visual and audio effects.

Like much or our work, this project is a continuation of an ongoing inquiry. It is an ongoing research into materials and technologies as well as a re-examination of known conditions, accepted norms and established methods. This has lead us to an innovative solution and stimulating new way of approaching interior architecture. Without predefining architecture, we responded directly and intuitively to the material qualities of place.

The context and program for the production studio suggests an experience ordered like a film or freeway, framing and containing reality. The design engages the user, heightens their sense of awareness, and brings a deeper understanding and vitality to their experience.

The design examines the tension between materials, form and experience. The interior can be viewed as "a skin or surface wrapper that moves in and out alternately concealing and revealing the building fabric." The layering and sculpting of the newly formed surfaces weave together disparate and contrasting materials. Recalling film director Alfred Hitchcock's interest in openings as metaphors, here, too, voids are as important as surfaces, revealing an earlier pattern of materials or use.

Of particular interest is the idea of transcending traditional craft and elevating humble materials without trying to make them into something other than what they really are. It is an attempt to find and reveal the

试。这个探索鼓励使用者更加深刻和广泛地理解基本原理，以及存在于他们自身、自然世界的核心资源和人类共同文化之间的微妙关系。

木头和塑料两种一般用于表面的基本材料被用于空间的塑造。100英尺长的木墙是由一个直接的转换方式而形成。木墙的电脑模型被直接送到由电脑自动控制的刨床上，74块厚度不一的胶合板通过自动控制设备雕刻成型，这在事实上取代了传统的手工方法。工作室的几个入口门被无痕迹地嵌入木墙纹路中。木墙表面具有的强烈的纵深空间旋转感，它充满无限活力，使用者和来访者穿梭其中，木头本身所具有的呆板感被消除了。

与木墙的建造工艺相对照，1/8英寸厚的带色有机板被贴覆在1英寸厚的办公室正面。嵌板被位于办公室内的顶灯从里面照亮，并转化为一种具有巨大空间深度和色彩感的材料。光线和人的运动使得整个空间充满活力，创造了高质量的时间和运动。光线也作为设计策略，将你拉入并穿越整个空间，它是时间隧道的记录器和社会连接器。这种有机板也像木墙一样展示了平凡事物中的非凡涵义。

在《建筑的复杂性和矛盾性》一书中，罗伯特·文丘里（Robert Venturi）这样写到："一个原本为我们所熟悉的事物，经由新的方法改造后，会带给人们新老并存的感觉。"把物体和材料放到"框架以外"，一个新的框架就能加深我们的感知。艺术不会复制我们所看到的东西，而会让我们看到新的东西。

extraordinary from within the ordinary. The exploration encourages the user to forge a deeper and more meaningful understanding of the fundamental, yet delicate relationships that exist between themselves, the natural world, its vital resources, and our collective cultures.

Two basic materials, wood and plastic are transformed from benign surfaces into sculpted space. The one hundred foot long wood wall was created by a direct transfer method. Computer models were sent directly from the architect to a computerized CNC router where 74 Glue laminated beams of varying thickness were sculpted by direct automation, virtually eliminating the traditional handcraft. Several studio entry doors were integrated into the pattern of the wood and seamlessly disappear. The result is a surface that is spatial, has depth and comes alive with movement. The perception that wood is a static dead material is ransacked. It is, in fact, alive with energy and moves through the space with its occupants and visitors.

In contrast to the carving method of the wood construction, 1/8" colored acrylic panels were layered to a thickness of 1" for the facades of the adjacent lead lined offices. The panels are backlit from large skylights located within the interior of their respective offices and are transformed into a material of considerable spatial dept and color. The movement of light and people engages and activates the entire

Opposite: wooden wall construction detail

木墙的施工及细部

Right: rendering of the acrylic panels

有机板效果图

space, creating a quality of time and movement. Light is also used as an ordering device: drawing you into and through the space. It is a register of the passage of time as well as a social connector. As in the construction of the wood wall the acrylic panels reveal the extraordinary from within something very ordinary.

In "Complexity and Contradiction in Architecture", Robert Venturi writes, "A familiar thing seen in an unfamiliar context can become perceptually new as well as old." By placing objects and materials "outside the frame," a new frame of reference deepens our sense of perception. Art does not reproduce what we see; rather it makes us see.

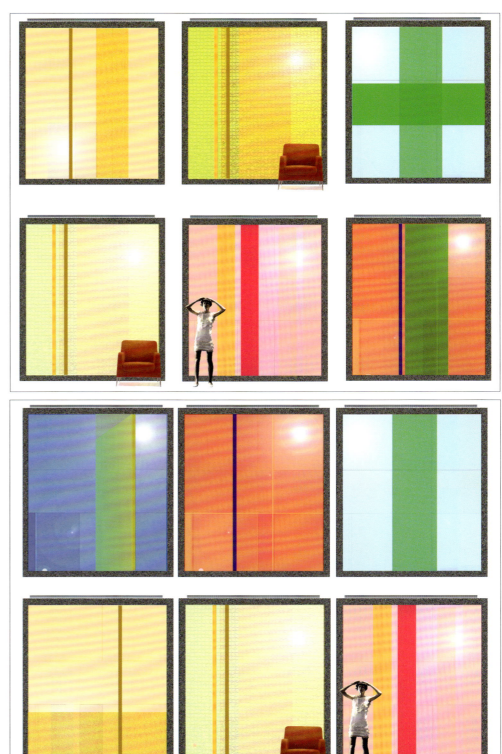

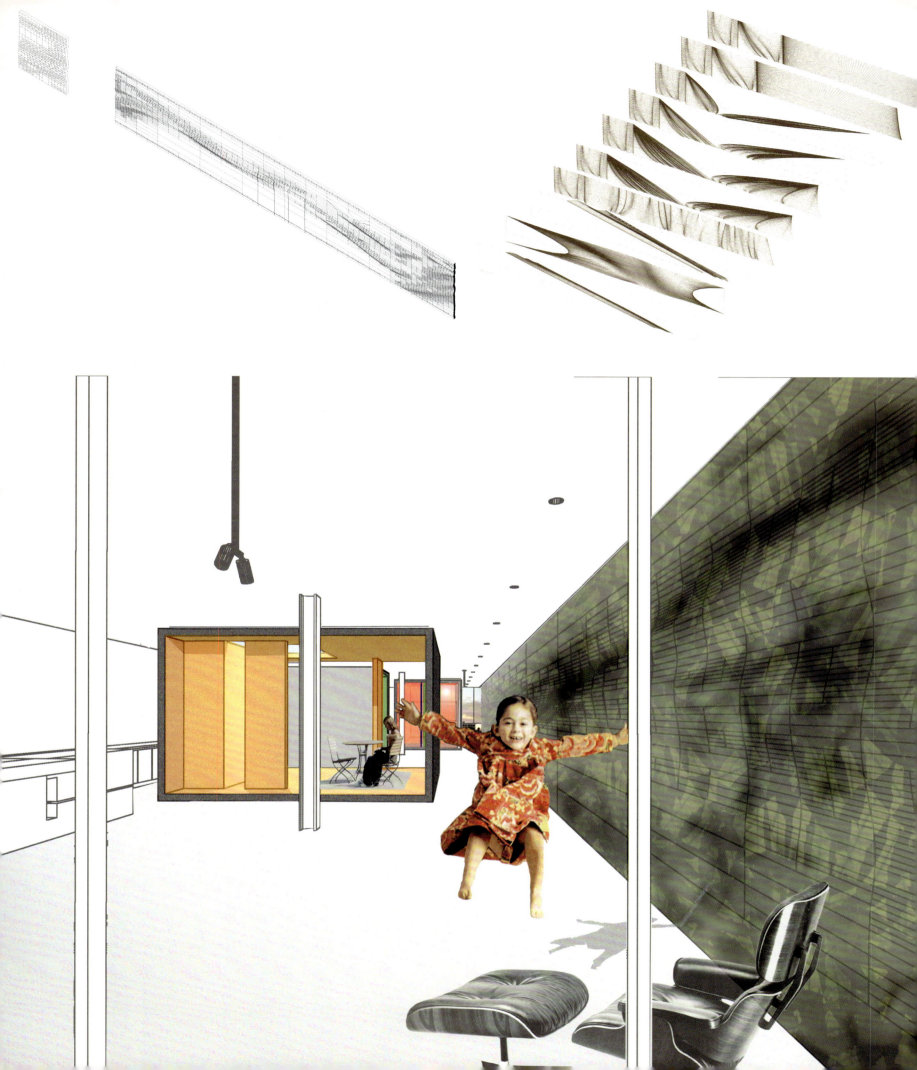

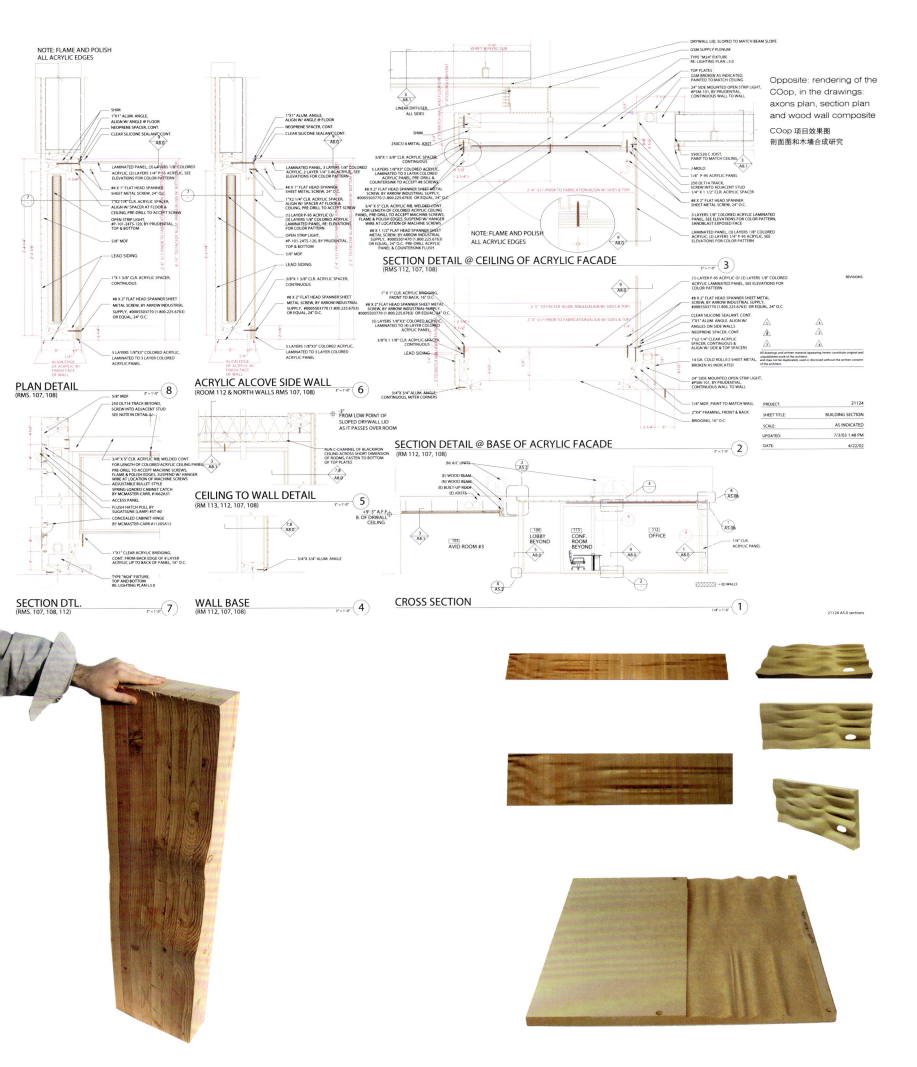

Opposite: rendering of the COop, in the drawings: axons plan, section plan and wood wall composite

COop 项目效果图
剖面图和木墙合成研究

Opposite: interior view of COop, the wood wall detail
COop 室内和木墙细部

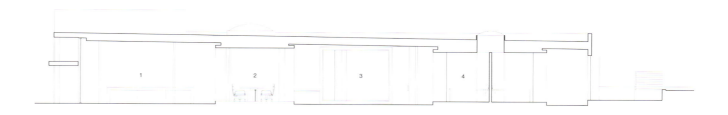

Left and right: interior view and the section plan of the COop

COop 项目室内和剖面图

1 LOBBY
2 CONFERENCE ROOM
3 OFFICE
4 RESTROOM

Contents

Introduction with Article Review, Page 7. **Project Survey** Colorado Court Page 23. COop Editorial Page 41. **Nascent Terrain** Page 59. Diva Page 67. Orange Grove Page 77. North Point Page 85. Jigsaw Page 95. Solar Umbrella Page 109. SM College Student Service Center Page 117. Bergamot Loft Page 125. Vail Grant House Page 139. The Firm Page 149. XAP Page 159. Reactor Films Page 171. Chronology Page 181 & Bibliography

Nascent Terrain

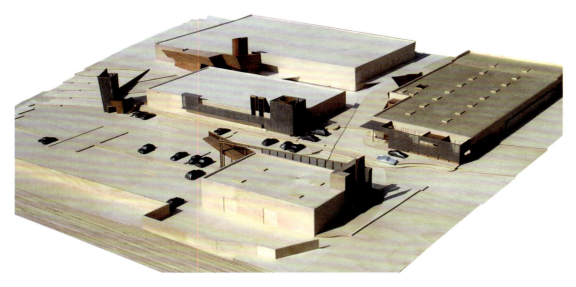

Nascent Terrain

Location of Project: 10119 Jefferson Blvd., Culver City, California
Client/Owner: DMC Investments
Total Square Footage: 70,000 sq. ft.

Project team: Lawrence Scarpa - Principal- in-Charge. Angela Brooks and Clay Holden – Project Architects. Kelly Bair, Peter Borrego, Angela Brooks, Anne Burke, Michael Hannah, Vanessa Hardy, Anne Marie Kaufman Brunner, Charlie Morgan, Fredrik Niilsen, Gwynne Pugh, Lawrence Scarpa.

General Contractor: Brad Brown/DMC Investments
Photography: Marvin Rand

The project seized as an opportunity to develop and foster social space in a city characterized by its desperate dearth of such space. The program involved creating a master plan for an existing industrial site comprised of a series of disparate warehouse buildings and surface parking. The program called for the conversion of 91,000 square feet of industrial brick building space into 70,000 square feet of creative office space and a small cafe. The existing site is bordered by Jefferson Blvd. and Ballona Creek in Culver City.

With the resolute intention of revitalizing an urban space, designs evolved to

生命之源项目

卡尔弗市，加利福尼亚州

这个项目抓住了一个在城市中发展和培养极度匮乏的社会空间的机会，并对一个拥有多栋仓库和地面停车场的工业社区作了新的建筑规划。项目计划把一个91 000平方英尺的砖混结构的工业建筑空间改造为一个70 000平方英尺的办公室和一个小咖啡馆。项目原位置与卡尔弗城市的杰斐逊大道和巴罗那湾接壤。

以复兴城市空间为目的，项目计划把一系列普通的工业建筑改建成为一个动态的、具有视觉和社会聚集空间双重功能的建筑。该项目围绕着"绿色社区"的概念创建了一个和周边建筑景点相连接的中心庭院。新的建筑景点成为社区中的地标。一部分老建筑被拆除并建成这个新的中心，新中心拥有充足的停车场和开放空间。这个复合的中心区被巧妙地设计成一个外部的庭院，新的地形特点定义了空间的聚合功能。一个水渠穿过庭院并和社区的其他部分相连，庭院中还有一个小巧的咖啡厅。

这个91 000平方英尺的老建筑的基本外观和结构核心在改建中被保留下来，并且更新了设施和服务设备。这些空间被改建成备用的办公空间，让未来的使用者能最灵活地按照需要对空间进行再安排。原建筑的外观和地形条件是项目改造的重点。地形条件被战略性地选择为社区建筑性改造的切入点。

Above: construction details of the Nascent Terrain

生命之源项目施工细部

Nascent Terrain

Below: modeling of the Nascent Terrain
生命之源项目模型

transform the series of undistinguished industrial buildings into a dynamic composition that would contribute to the urban-scape both visually and as a social gathering space. This project is organized around the concept of a " Campus Green." This "Campus Green" spatially connects the existing buildings by creating a central courtyard and architectural landscape. The new landscape and site furniture act as markers to orient one in the campus. A portion of the existing site was demolished to create this new center, provide for more parking and create more open space. The center of the complex is designed as an exterior courtyard; the ground plan is manipulated so that tilting planes define space and provide for places to gather. A channel of water cuts through the courtyard and ties it into to the rest of the site. A small cafe opens onto the courtyard.

The 91,000 square feet of tenant improvements were approached as basic shell and core remodels that included stripping the spaces down to their most essential elements and updating infrastructure and services. In essence, these spaces were prepared as "creative office space" theoretically allowing future tenant(s) maximum flexibility to adapt the spaces as needed and/or desired.

Primary emphasis was given to modifying the exterior buildings and site conditions. Critical points of

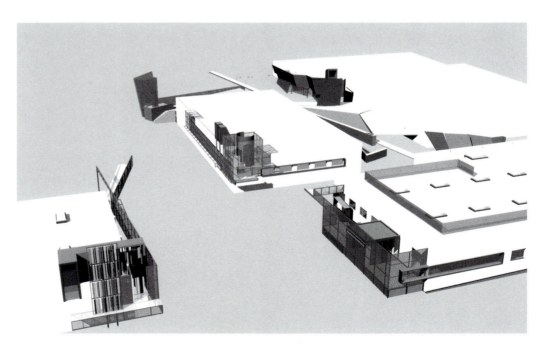

Above and below: modeling and rendering of the Nascent Terrain

生命之源项目效果图和模型

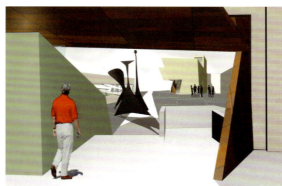

the corner of another building along Jefferson Blvd. and also frames the site entry. A tower folly completes the site and creates a gathering place and connecting link between two different levels. Even the surface parking is exploited as a site feature and woven into the overall composition.

Careful attention was paid to maximizing this project's potential for urban regeneration and the creation of a more interesting and enriching social space.

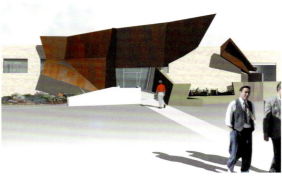

Above: renderings of the Nascent Terrain
生命之源效果图

interest were strategically chosen as sites that would receive architectural modifications. These sites were selected for their impact potential and manipulated to create thresholds, markers, visual interest and movement. Each intervention, or architectural face-lift, to a building was carefully considered and orchestrated with the site. Each building receives an illuminated steel screen to activate the adjacent space. A steel "butterfly wing" forms the entrance to one building and also ends the courtyard. A two-story steel "wind chime" grows out of

入口、路标、视觉景观和走向都依据地形特点而被巧妙的设置。每一次建筑外观的修改和翻新都经过认真的考虑，以保证和地形的风格合为一体。每一栋建筑都安装了明亮的钢屏幕而使得周围的空间更加有活力。一个钢制的"蝴蝶翅膀"状的物体形成了建筑的入口和庭院的出口。在杰斐逊大街旁一个建筑的角落中升出的一个两层的"风铃"构成了社区的入口。社区尽头的叠立塔形成了一个聚集地，也是两边建筑的连接点。甚至地面停车场都被开发成一个特色景点而被和谐地融入整个建筑群中。

这个项目最大限度地挖掘出了城市改造和丰富充实社区空间的潜力。

Left: construction details of the Nascent Terrain

生命之源项目施工现场

Nascent Terrain

Below: renderings of the Nascent Terrain
生命之源项目效果图

Contents

Introduction with Article Review, Page 7. Project Survey Colorado Court Page 23. COop Editorial Page 41. Nascent Terrain Page 59. **Diva** Page 67. Orange Grove Page 77. North Point Page 85. Jigsaw Page 95. Solar Umbrella Page 109. SM College Student Service Center Page 117. Bergamot Loft Page 125. Vail Grant House Page 139. The Firm Page 149. XAP Page 159. Reactor Films Page 171. Chronology Page 181 & Bibliography, Page 191.

Diva

迪瓦工作室
洛杉矶市，加利福尼亚州

这是一个对在20世纪40年代建成的零售大楼进行完全改建的建筑项目。3个现有零售店被改建为一个模拟法院和一个紧邻的为瑞典迪瓦公司建造的小型舞蹈工作室。这个项目包括一个设有视频的会议室和一个双向观摹的模拟法庭。所有的内部家具都是由皮尤+斯卡帕设计的。

这个项目的特点是设计的二元性。在同一个空间中经营两个截然不同的业务。建筑的一边是"完全"诉讼发布机构——一个训练法院书记官的模拟法庭环境。另一边是瑞典迪瓦制作公司的舞蹈排练工作室，这是一个为私人聚会和活动服务的表演组织。整个建筑为两个组织同时诠释了各自的特点。

每个空间都被有效地规划和设计，从而显现与邻居不同的元素。在空间的中央，有一堵墙划分着"完全"诉讼发布机构和瑞典迪瓦制作公司的舞蹈工作室。两个机构的接待区、办公室和供应空间都沿着这面墙布局。一面单面镜墙将双方隔开，舞蹈工作室的表演者能在工作中窥探模拟法庭中的活动，法官们的调查和取证都在不知不觉中成为被观察的对象。隐藏在灯光设备和法庭外观中的音像视频设备是评判教学过程的工具，它们进一步挑战着成为被监督对象的法律专家们。

这个建筑的重要特点也在它的正面外观中得以体现。一个玻璃和铝制表面为"完全"诉讼发布机构创造了一个透明的外观，一种波状金属板和水泥墙为瑞典迪瓦制作公司的舞蹈工作室创造了不透明的内部工作环境。整个设计掩饰了两个空间的现实关

Diva

Location of Project: Los Angeles, California
Client/Owner: Leanna Green
Total Square Footage: 3100 sq. ft.

Project team: Lawrence Scarpa and Gwynne Pugh - Principals - in-Charge, Peter Borrego, Angela Brooks, Heather Duncan, Charlie Morgan.

Structural Engineering: Gwynne Pugh
General Contractors: CWE Construction
Photography: Benny Chan - Fotoworks

Program: A complete remodel and new storefront facade for an existing 1940's retail building. The three existing retail stores were converted into a training facility for court reporting and an adjacent small dance studio for Swedish Diva Productions. The project includes a state of the art mock training courtroom with video conferencing capabilities and two-way viewing. All interior furnishings are custom designed by Pugh + Scarpa.

Solution: The architectural response to this project emerged from the unique duality of its program. Contained within one shell are two disparate businesses managed by a singular proprietor. One side of the building is occupied by Absolute Court Reporting, an establishment that trains court reporters in a state of the art simulated courtroom environment. Directly adjacent within the same shell resides the dance rehearsal studio for Swedish Diva Productions, a performance group catering to private parties and events. The architecture directly addresses this juxtaposition in both its organization and formal articulation.

Each space was strategically planned so that the most critical programmatic element of one would sit directly astride that of its neighbor. Thus, sharing a dividing wall at the core of the space are the mock courtroom of Absolute Court Reporting and the dance studio of Swedish Diva Productions. Reception areas, offices and support spaces unfold along the respective perimeter walls of each business. A one-way mirrored wall separating the courtroom and dance studio subverts the relationship between viewer and viewed. Performers accustomed to exhibiting themselves in their work become voyeurs gaining privileged views into the activities of the mock courtroom. Participants in the mock courtroom unknowingly become viewed object—their role as information scrutinizer, inquisitor and truth seeker elegantly subverted. While considered critical to the teaching process, state of the art video and audio monitoring devices concealed in the lighting fixtures and surfaces of the courtroom further challenge the role of the legal professional who becomes the surveilled object.

系，路过者绝不会想到两者是在共享着同一个核心空间。一个外露的缠绕着整个建筑的管状钢支架是两个空间惟一的物理联系。这个管状钢支架为两个不同的机构提供了结构支撑。

该项目的其他特色包括："完全"诉讼发布机构的一个曲线形透明金属装饰墙，这堵墙定义了一个接待区和工作站并在视觉上和物理上引导使用者沿着走廊通向一个厨房和服务区。另外，一些专门的设计也应用在模拟法庭、细木家具和办公用品中。

The critical content of this project's architectural response is also evidenced in its public facade. A glass and aluminum storefront creates a transparent condition for Absolute Court Reporting while a corrugated sheet metal and concrete wall create an opacity that blocks out any potential voyeuristic views into Swedish Diva Productions. The formal resolution of the facade belies the reality of the relationship between these two spaces. A passer by would never know that they were actually intimately connected and sharing the same core space. An exposed tubular steel support spanning both spaces provides the only overt physical clue that there is any connection between the two operations. Significantly, this tubular steel element provides structural support for both businesses identifying signage.

Other distinct features of this project include a canted translucent metal stud wall that defines a reception area and work stations for Absolute Court Reporting and leads the user visually and physically down the main circulation corridor to a kitchen and service area at the rear of the space. Additionally, custom designs were provided for the mock courtroom furniture and much of the built in cabinetry and office furniture.

copy center
office support staff
court/mock trial room
kitchen
deliberation room
dance studio/courtroom viewing

COURT

DANCE STUDIO

court entry
dance studio entry

Opposite: interior view of the Absolute Court Reporting house and the section plan at storefront

"完全"诉讼发布机构室内和店面剖面图

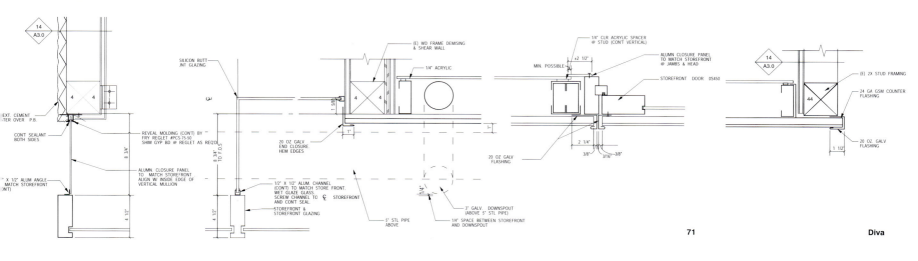

Left: exterior detail of the Absolute Court Reporting house

"完全"诉讼发布机构室外细部

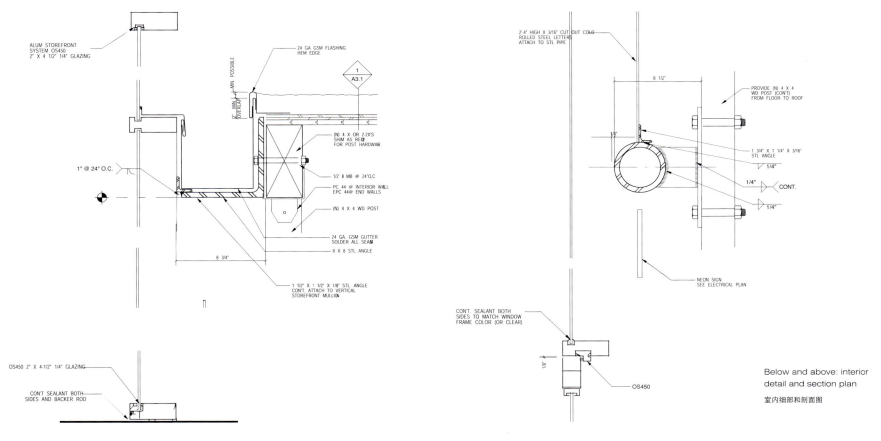

Below and above: interior detail and section plan

室内细部和剖面图

Contents

Introduction with Article Review, Page 7.

Project Survey

Colorado Court Page 28. COop Editorial Page 41. Nascent Terrain Page 59. Diva Page 67. **Orange Grove** Page 77. North Point Page 85. Jigsaw Page 95. Solar Umbrella Page 109. SM College Student Service Center Page 117. Bergamot Loft Page 125. Vail Grant House Page 139. The Firm Page 149. XAP Page 159. Reactor Films Page 171.

Chronology Page 181 & Bibliography.

Orange Grove

橘林住宅

西好莱坞，加利福尼亚州

该项目位于传统的平房式单栋家庭住宅区，是西好莱坞区一个新地标。建筑设计得很感性并且与周围的住宅环境十分协调，但是又与这些住宅在材料和建筑格局上不尽相同。这个住宅借鉴了现代主义的建筑形式，没有采用传统的斜屋顶。橘林住宅电子一体化特色是西好莱坞非传统建筑的代表。它与周边的建筑结构存在明显的不同，住宅由巨大的非常有用的正面阳台区与街道产生了紧密的关系。

虽然一些大型的、戏剧化的布局元素定义了这座建筑，但同时也着眼于人类所能把握和理解的视觉尺度，住宅本身也分为两个不同的部分。它与加利福尼亚州现代主义的标志建筑——申德勒住宅的短距离分离模式类似。在申德勒住宅中，窗户和走廊这些传统建筑元素成为抽象雕塑集合体的一部分。申德勒住宅的窗户安置在结构混凝土墙板之间的间隔中。橘林住宅的窗户也被插入建筑不同部分之间的间隔中。

橘林住宅的设计通过一种微妙的张力平衡而体现。建筑体积、窗户的安置、门和阳台都不是静止的，而是构成了一个运动的、活跃的三维混合体。建筑中的每一个小部分都拥有强大而清晰的形状，例如东面和北面二楼阳台的波状金属面罩。另一个清晰的例子是建筑正面两个一大一小的正方形阳台之间的对话：一个是开放式的，另一个则被不锈钢条面罩所遮掩，两个阳台都能在其他建筑元素之间起到平衡作用。小阳台通向下面的车库门，大阳台通往卷帘门和第一层的阳台。每一个建筑元素都可以解读为隐含于它们自身的抽象形式——比如窗户可以解读为建筑的缝隙，或者可以解读为加框的盒子，

Orange Grove

Location of Project: West Hollywood, California
Client/Owner: Chris DeBolt
Total Square Footage: 6,700 sq. ft.

Project team: Lawrence Scarpa, AIA - Principal- in-Charge. Angela Brooks, AIA, Silke Clemens, Vanessa Hardy, Ching Luk, Tim Petersen, Gwynne Pugh, AIA, Lawrence Scarpa, Katrin Terstegen.

Engineering: Oxford Engineering
General Contractor: Becker General Contractors
Photography: Marvin Rand

Located in a neighborhood characterized by traditional bungalow style single family residences, Orange Grove is a new landmark for the City of West Hollywood. The building is sensitively designed and compatible with the neighborhood, but differs in material palette and scale from its neighbors. Referencing architectural conventions of modernism rather than the pitched roof forms of traditional domesticity, the project presents a characteristic that is consistent with the eclectic and often unconventional demographic of West Hollywood. Distinct from neighboring structures, the building creates a strong relationship to the street by virtue of its large amount of highly usable balcony area in the front facade.

While there are dramatic and larger scale elements that define the building, it is also broken down into comprehensible human scale parts, and is itself broken down into two different buildings. Orange Grove displays a similar kind of iconoclasm as the Schindler House, an icon of California modernism, located a short distance away. Like the Schindler House, the conventional architectural elements of windows and porches become part of an abstract sculptural ensemble. At the Schindler House, windows are found in the gaps between structural concrete wall panels. At Orange Grove, windows are inserted in gaps between different sections of the building.

The design of Orange Grove is generated by a subtle balance of tensions. Building volumes and the placement of windows, doors and balconies are not static but rather constitute an active three-dimensional composition in motion. Each piece of the building is a strong and clearly defined shape, such as the corrugated metal surround that encloses the second story balcony in the east and north facades. Another example of this clear delineation is the use of two square profile balcony surrounds in the front facade that set up a dialogue between them—one is small, the other large, one is open at the front, the other is veiled with stainless steel slats. At the same time each balcony is

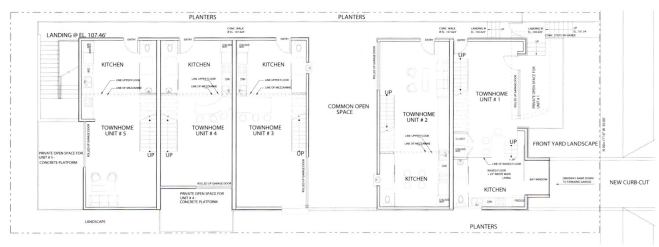

Ground Floor Plan

Below and Opposite: rendering of Orange Grove ground floor plan and second floor plan

橘林住宅效果图，底层和二层平面图

Second Floor Plan

81 Orange Grove

Opposite and below:
rendering of Orange Grove

橘林住宅效果图

balanced and related to other elements in the building, the smaller one to the driveway gate below and the other to the roll-up door and first floor balcony. Each building element is intended to read as an abstract form in itself- such as a window becoming a slit or windows becoming a framed box, while also becoming part of a larger whole. Although this building may not mirror the status quo it answers to the desires of consumers in a burgeoning niche market who want large, simple interior volumes of space, and a paradigm based on space, light and industrial materials of the loft rather than the bungalow.

同时它们也是建筑整体的一部分。虽然这个建筑并不能反映当前市场的现状，但是它回应了一个日见增长的消费市场，消费者越来越需要一种具有大且简单的内部空间格局，以及一种基于空间、光线和工业材料的阁楼式样板而非平房式的建筑。

Below: rendering of Orange Grove
橘林住宅效果图

Contents

Introduction with Article Review, Page 7.
Project Survey Colorado
Court Page 23. COop Editorial Page 41. Nascent
Terrain Page 59. Diva Page 67. Orange Grove Page 77.
North Point Page 85. Jigsaw Page 95. Solar
Umbrella Page 109. SM College Student
Service Center Page 117. Bergamot Loft Page 125.
Vail Grant House Page 139. The Firm Page 149.
XAP Page 159. Reactor Films Page 171.
Chronology Page 181. & Bibliography

North Point

北点住宅项目
波士顿剑桥，马萨诸塞州

北点社区项目提供了一个良机让我们以一个全新的视角来审视社区建设。现代都市可以提供风格迥异的多层次的社区感受——从公众到个人，从社区到家庭。

在过去，鲜有设计师在规模宏大的住宅项目中尝试都市化社区空间的家庭化建筑布局。比如波士顿巴克湾（Back Bay）历来的特色建筑都是5层以下的联排房。一部分这类建筑已被改造用于商用。不过联排房的主要形式一直还保留着。相反，在普鲁登肖中心（Prudential Center）的住宅的现代结构特点是由一个大厅控制的多重复式公寓建筑，这些住宅超越了仅仅从私人内部空间来审视城市关系的建筑。没有人尝试把这两种相反的模式结合起来并开发两者能给予对方的内涵，其主要的原因来自于复杂建筑布局的巨大挑战，以及开支、结构和机械运作效率等问题。

北点社区示范了都市社区住宅的创新模式，其中包括4种不同的在对方基础上形成的新类型：城市阁楼（内墙隔断的开敞式平面布置住宅）、二层无电梯公寓类型、有门廊花园的联排房类型和高层单双走廊公寓类型。

这个计划有效地解决了管槽的堆放排列、隔断和其他设施间的协调、框架和半地下车库结构的统一性问题。总之，在达到建筑布局丰富多样性的同时，也保证了项目在开支和材料使用上的高效和节省。

这种设计方法使项目达到了几个重要目标：

North Point

Location of Project: Cambridge, Massachusetts
Client/Owner: Spaulding and Slye/ Colliers International
Total Square Footage: 366,000 sq. ft.

Project team: Office dA – Nader Tehrani, Monica Ponce de Leon – Principals - in-Charge Julian Palacio, Michael Tunkey – Project Coordinators, Sean Baccei, Ghazel Abbassi, Penn Ruderman, Krists Karklins, Ehtan Kushner, Lisa Huang, Elane Chow

Project team: Pugh + Scarpa - Lawrence Scarpa, AIA - Principal- in-Charge. Angela Brooks, AIA, Silke Clemens, Michael Hannah, Vanessa Hardy, Ching Luk, Fredrik Nilsson, Gwynne Pugh, AIA, Lawrence Scarpa, Katrin Terstegen .

Engineering: DM Berg, Structural
OVE Arup, Mechanical, Electrical and Plumbing
Cost Estimating: Rider Hunt Levett & Bailey

North Point: Pattering a New District

The North Point site offers an opportunity to develop a new community and unique urban fabric. The challenge of the project is to scale the buildings so that its inhabitants may identify themselves with the various "degrees" of community that the modern metropolis has the potential of offering –from the public to the private, and from the communal and individual.

Traditionally, there are few precedents that take on the challenge of working at both the domestic and urban scale in large housing projects of some programmatic ambition. For example, in Boston's Back Bay, the historic core is predominantly characterized by row houses of five stories and less. In few instances these structures have been transformed to easily adapt for commercial or retail uses at their base, but the row house remains a relatively stubborn to major transformation. Alternatively, modern structures such as the residential buildings in the Prudential Center are characterized by repetitive apartment blocks that are controlled by a lobby and have little relationship with the city beyond the views from the private interior spaces. There are few examples of buildings that attempt to bring together these two conflicting models and exploit the richness of what each has to offer in combination with the other. There is a good reason why there are few precedents for this proposed typology for urban housing. It is a considerable challenge to produce programmatic richness and varied housing types, while also achieving the appropriate economic, structural and mechanical efficiencies that is characteristic of this kind of development.

Below: rendering of North Point

北点项目效果图

Our proposal for North point demonstrates an invention for urban housing that includes four differing typologies that are stacked intricately on top of each other to create a new hybrid: urban loft/work-live spaces at street and plinth level, row house duplex types with stoops and gardens at street and plinth level, NYC style walk-ups on the second floor, and double and single-loaded corridor apartments at upper levels.

Our proposal also efficiently solves the basic alignment of stacked plumbing chases, coordination of the structural bays with a variety of differing stacked unit plans, and the synchronization of the structural grid with the semi-subterranean garage. In summary, the programmatic richness and unique variation is achieved while maintaining an efficient and lean structure, both economically and well as material usage.

This type of approach achieves several important goals for this project:

- It maximizes street life by making as many unit types as possible directly accessible to the sidewalk, making a richer urban lifestyle. The lofts, townhouses, and the walk-ups, as well as the public access points to the apartment building all prominently address the city and adjacent street life.
- It produces a wider variety of units, offering a broader set of choices to a wider demographic profile,

North Point

——通过建造不同的住宅单元类型求得最大化的街区生活，并创造一种更丰富的都市生活方式。阁楼、双层公寓、无电梯的公寓以及公共社区，这些都是都市和街区生活的彰显。

——种类广泛的住宅单元类型，可提供给更广泛的人群，使人们有更多样的生活方式和经济选择。相应地，也使得该建筑更加市场化和经济适用。建筑布局的灵活性很强，今天的阁楼也许很好，但是在将来也很容易改建成商用空间。

——每一种类型的房屋都会提供和它们居住区相关的风格：法式门/朱丽叶阳台、传统阳台、双倍加高生活平台、院子、门廊和屋顶花园，所有这些在居住者和城市之间维系了一种独特的关系。

——该建筑在不缩小项目规模以模仿原有环境的情况下提供了不同规格的住宅。

——创造了一系列的在个体单元和城市格局中的缓冲公共空间。

——该设计提高了建筑的采光和通风能力，并最大限度地获得了自然采光和通风，由此获得了良好的室内舒适度。

都市主义：建筑与庭院的协调

个体单元、住宅类型、社区和城市这四个都市空间相互协调。S和T形场地由一个我们叫做南部公园的城市花园连接在一起。南部公园由不同的"细条形"的硬质和软质景观组成，在地面与S和T形场地相连。建筑周围的一排绿树穿过查尔斯·史密斯综合性建筑，并加强了与位于南北中轴线上的东剑桥和中央公园之间的联系。

T形场地由两个庭院构成。一个是位于+4′-0″的半公共冬季花园，另一个是位于+30′-8″的半私人化夏季花园，其间连着温泉、桑拿浴和其他公共

diverging lifestyles, and economical choices. In turn, this makes the building more marketable and economically viable. The open plans of the base units are designed with a maximum flexibility, working well as lofts today, but easily transformable to retail of commercial spaces in a future life.

- Each unit offers different kinds (and dimensions) of open space in association with its living areas: French door/Juliet balconies, conventional balconies, double height living decks, yards and stoops, roof gardens, all of which offer a distinct relationship between its inhabitants and the unique perspectives they offer towards the city.
- It offers a variety of scales for the buildings, without miniaturizing the scale of the project to mimic the historic context.
- It produces several kinds of public space that mediate between the scale of the individual unit and the scale of the city.
- It enhances the ability to achieve maximum natural day-lighting, ventilation, and indoor are quality.

Urbanism: Mediation with Courtyards

Four urban spaces help to mediate the relationship between the individual units, housing types, blocks, and the city. Site S and T are held together by a discreet urban garden we call South Park. South Park is composed of various "stripings" of hard and soft landscapes, connection the 'S' and 'T' sites on the ground level, while an allee of trees reinforces the broader site's connection to East Cambridge and Central Park on the north-south axis through the Charles Smith complex.

Site T is organized around two courtyards, one of which is semi-public in nature at +4'-0" (the Winter Garden) and the other which is semi-private located at +30'-8" (the Summer Garden) in relationship to the spa, sauna and other public amenities of the building. While the Winter Court enjoys a healthy dose of sunshine, it is protected with trees, bamboo gardens, decks, and stoops, providing for a smaller community of duplexes, walk-ups and lofts to share a common ground for relaxation, gardening, and conversation. Serving the entire building, the Summer Garden enjoys an elevated and privileged view of Central Park; shaded by the building, the space benefits from the cool breezes of the summer and provides some outdoor space for exercise, yoga, and relaxation.

Site S is organized around a third court (the Community Court) whose distinct interiority is to create and common space for events, gatherings, and social occasions. Protected from the street, it is also linked to Central and South Park through vertical cracks that help to create cross ventilation through the space. A Community Space is located next to the courtyard and opens into it for events that require indoor and outdoor activities.

The spa on the second and third levels overlooks the courtyard and has exterior stairs and balconies tying together the levels. At the 4th, 6th and 8th floors bridges cut diagonally across the court space; these bridges are conceived as open seasonal bridges. The bridges transform a typical interior corridor into an exterior balcony which allows the user to visually participate in the court space. They also provide shading to the court. The facade is further animated by window boxes and planes articulated as volumes unto themselves, which push and pull from the facades primary folds. The formal geometries and material richness of this facade have a dynamic effect on the in between courtyard space that now flourishes as a kind of piazza for itself and the surrounding building.

Working from the Inside Out: A Room with a View

As a basic methodology, we have developed our unit types as a vehicle of optimizing the relationship between interior and exteriors spaces and their views, while developing a strategy for fenestration and the massing of the buildings.

The live/work spaces at the base of the buildings are organized as split levels, so that the units can be used flexibly –the levels that are flush with the sidewalk being used as work spaces, while the upper levels that are at the Garden level serving as the private areas of the living section. Some of these units have direct access into the parking below. The base of the building is treated as civic storefront conditions, whose interiors can be selectively exposed or protected based on their use by its inhabitants—through frosting, curtains, scrims or blinds. By organizing this in a split-level condition, we create a 16 foot high space, which adds to the feeling of openness in each unit.

The townhouse duplexes are accessible to the street by way of stoops, as is customary in other Boston neighborhoods. Some also have access to the Winter Garden on the T site, and the Community Court on the S site, while also having direct access to the garage as walk ups.

The buildings' fenestration is planned to correspond to the various patterns of individual unit plans. The single bay vertical window corresponds to smaller rooms and its verticality establishes a rapport with the human body and totemic objects in the city skyline. The double bay vertical window for medium sized rooms (the French door Juliet balcony) operates as a hybrid, creating a flexible relationship between the indoors and outdoors. The ribbon window, alternating between larger and smaller rooms, establishes a more distinct rapport with the horizon and the skyline as framed the longitudinal opening. At building corners, the units have folding glass walls that open completely, eradicating the boundary

娱乐休闲设施。在冬季庭院里，人们能够享受充足的阳光，那里有绿树、竹园、平台和门廊，这里是小型的复式公寓、无电梯公寓和阁楼的综合社区，还是人们共同休闲、从事园艺和交谈的美妙地方。夏季花园也为整个建筑增色不少，从那里能够直接欣赏到中央公园的景色，夏季的凉风拂面而过，园外还有健身、瑜珈和放松的场所。

S形场地是社区庭院，这是人们进行聚会、社交和其他特别活动的地方。它通过空间中加强通风的垂直间隙连接着中央和南部公园。一个社区空间紧邻着庭院，里面可以进行室内和户外的各种活动。第二、三层设有可俯瞰庭院的温泉，相互间由外部楼梯和阳台相连。第四、六、八层公寓的连桥沿斜边跨越庭院空间，这些连桥仿佛是露天的观景桥；它们把一个典型的内部走廊转化成一个外部阳台，使使用者能够看见庭院空间中的活动。它们同时也为庭院提供遮挡。窗槛花箱和极具节奏感的表面起伏增加了建筑外表的活力。其外表的几何形态和丰富的材料是庭院空间中的动态表现，使庭院广场和周围的建筑显得华美繁荣。

由内到外的设计——看得见风景的房间

我们以房屋的类型为基点，并把它们开发成一种优化内外空间相互视点关系的工具，同时也依据此发展了开窗法和建筑分布格局方面的策略。

建筑的生活和工作空间被分开，这样一来，各个单元可以被灵活地使用。有走道的一层被用作工作空间。连着花园的上半层是私人生活空间。其中一些单元能够直接通往下面的停车场。建筑的底层基本被设计为城市店面形态，内部的空间会根据居住者的使用需要布置成无掩蔽状态，或者通过

Opposite: interior rendering of the North Point

北点室内效果图

between the interior and exterior. At the street and courtyard levels, curtain wall and storefront conditions establish a more direct and privileged relationship with either the sidewalk or gardens. On the upper levels, where privacy is never compromised by proximity to other units, curtain walls provide for amplified views of the park and natural light. The resulting effect is a project that possesses great variety, richness of place essential to human habitation and quality of life.

The Building Circulation:
The path of travel to building and unit entries are enhanced by corridors that have views and natural light. Whether single or double loaded, the corridors are a pleasant transition from the entry to the individual units.

Sustainable Strategies:
Our team is recognized as an expert in the field. We are able to provide long and short term cost benefit analysis, DOE 2 energy modeling, and building commissioning. We propose to explore options for discussion and possible implementation; our basic strategy is as follows:

Heating and Cooling:
Each apartment will feature exceptional air quality, natural light and ventilation. Every unit will be served by its own water source heat pump unit which will be capable of providing both heating and cooling as necessary. When heating is required, the heat pump will draw thermal energy from a condenser water circulating system and discharge it into the apartment. When cooling is required, the heat pump will reverse and draw thermal energy from the apartment and reject it to the condenser water loop. At a building level, if the condenser water has a net heat deficit, it will require heating up. Similarly, if most apartments are in cooling mode, it will be necessary to remove heat from the condenser water circulating loop. It is proposed that this thermal regulation is provided by a ground water source heat pump (a geothermal system) that will either draw energy from the ground water, or reject heat to the ground water, depending on building needs. Such a system will maximize energy efficiency while keeping the maintenance of central systems to a minimum—unlike a central boiler/chiller system that is more maintenance intensive and likely to be more costly to install.

A Green Roof:
Our proposal incorporates a green roof that creates a thermal mass separation and thermal break from the outdoor climate. The Green roof also converts CO_2 to Oxygen. It also provides for storm water control. In summary, the green roof provides for an energy efficient building and a reduction of CO_2 for the city.

Photovoltaic Solar System:
The photo voltaic system is designed to offset the cost of utility usage for the common areas which is traditionally paid for and maintained by the building owners or homeowners' association.

窗帘等遮挡物保护起来。我们创造了一个16英尺高的空间来达到这种分层模式，也加强了每一个单元的开阔感。

复式联排住宅单元可以穿过门廊通往街道，这是波士顿常见的住宅模式。其中一些还可以通往T形场地的冬季花园和S形场地的社区庭院，同时可以直接走下车库。

建筑开窗法按照不同类型的住宅单元平面来设计。竖向单凹窗适用于较小的房间，它的垂直感在人和城市天际线之间建立了某种联系。竖向双凹窗适用于中等面积的房间，是可以混合操作的连体法式门和朱丽叶阳台，体现了内外空间的灵活性。在较大和较小的房间之间交替穿插的带状窗户是观看地平线和城市景观的取景框。在窗边的内角安装有能够完全打开的折叠玻璃墙，这是内外空间的分界线。在街道和庭院一层，其间的幕墙和店面形态更直接地确立了人行道和花园间的优先关系。在建筑上层，住户的私密并没有因建筑格局的紧密性而受到影响，幕墙提供了更广阔的鸟瞰公园的视野和享受阳光的机会。整个项目为居住者提供了丰富多样的生活空间和优良的生活品质。

交通空间

可以观景而且采光充足的走廊是通往建筑和房间入口的通路。无论是单人还是双人通道，走过这些走廊到达各个私人房间的感受都是很美妙的。

可持续性策略

我们的团队在这个领域被视作专家。我们能够提供长、短期成本收益分析、DOE2能源模型和建筑施工委托。我们的可行性方案和基本策略如下：

供热和制冷

Opposite and below: rendering of North Point
北点项目效果图

The PV area calculation is as follows:
Area of circulation + corridors X $0.50/sf = cost of electricity
Cost of electricity ÷ $0.15/kWh = kWh of electricity
kWh ÷ 1500 hrs @ peak = kW output of PV's
kW output ÷ 8 w/sf + sf of PV's

Fenestration:
The North and South elevations are proposed to have LOW-E glass. The south facing windows would also be protected by horizontal visors and sun shades, protecting the interior. The interiors should be protected by white "Mecho" type shades, reflecting undesired solar gain. Also the shade system allows individual thermal control. The east and west elevations are proposed to have LOW IRON glass with vertically oriented white louvered shades on the interiors. The curtain wall portions have exteriors louvers also, protecting the facade from the extreme southern sun.

Other Green Strategies:
Each floor contains chutes for the collection of recycled materials. Indoor materials such as paint and carpets are proposed to be low V.O.C. and high in post industrial and post consumer materials.

每一套公寓都能获得优质的空气、采光和通风。每一套公寓都有独立的水循环冷热供应系统。当住户需要供热时，热水泵就会从一个冷凝器水循环系统中抽取热量并传送到公寓。当需要制冷的时候，热水泵就会抽走送往公寓的热量并送回冷凝器水循环系统。每层建筑会因相互间冷热使用的不均衡而造成建筑内的冷热水循环系统的工作障碍，这时一个地下水资源供热泵（一个地热系统）会根据住宅的需要从地下水提取或者释放能量，从而协调建筑内的冷、热水循环系统的工作。这样一个系统将最大程度地节省能源，并将中心系统的维护成本降到最低——不像传统的中央供热/制冷系统那样需要大量的维护工作和高昂的安装费用。

绿色屋顶
我们设计了一个能够将户外热量分流和终止的绿色屋顶。它能够将二氧化碳转化成氧气，还能够控制雨水。总之，绿色屋顶对节能建筑和减少城市二氧化碳不无裨益。

太阳能光电系统
太阳能光电系统能够弥补原先分摊在业主账户上的公共水电费。

太阳能光电系统（PV）面积计算如下：
(循环面积+走廊面积)x $0.50／平方英尺=电费
电费÷$ 0.15/kWh=用电量（kWh）
kWh÷1500hrs@ 最高值=PV电力输出量（kW）
PV电力输出量÷8w/ft²+PV平方英尺数

开窗法
建筑的南、北两面使用低辐射玻璃。南向的玻璃窗装有水平的白色"Mecho"遮阳篷保护室内，这种遮阳篷可反射紫外线和调节热流。东、西两面的窗户装有有色玻璃和内置百叶窗帘，幕墙外置百叶窗可抵挡由正墙反射来的阳光。

其他的绿色策略
建筑的每一层都设有垃圾回收站，室内使用的涂料和地毯都计划使用低V.O.C后工业材料。

Below: modeling of North Point
北点项目模型

Below: renderings of North Point

北点项目效果图

Contents

Introduction with Article Review, Page 7.
Project Survey Colorado Court Page 23. COop Editorial Page 41. Nascent Terrain Page 59. Diva Page 67. Orange Grove Page 77. North Point Page 85. **Jigsaw** Page 95. Solar Umbrella Page 109. SM College Student Service Center Page 117. Bergamot Loft Page 125. Vail Grant House Page 139. The Firm Page 149. XAP Page 159. Reactor Films Page 171. Chronology Page 181 & Bibliography.

Jigsaw

Jigsaw 公司
洛杉矶，加利福尼亚州

电影编辑住在科幻世界中，电脑屏幕就是他们生活的全部。为了完全投入到这个世界中去，像反光这样分散注意的事物都需要被排除掉。所以，他们的生活通常会处于一个密闭的黑匣子里。那么，建筑师们如何为他们设计一个充满刺激的，同时满足社交与隐居需要的工作环境呢？

面对这个挑战，皮尤+斯卡帕接受委托设计了Jigsaw电影剪辑公司。建筑师们把一个大约5000平方英尺的弓形拱顶仓库变成了一个崭新的不可想像的世界。Jigsaw 位于洛杉矶西区一个周围环境没什么特点的工业区。建筑以一种独立的形式把分离的内部空间和怀旧的外部都市结构结合起来，它对于洛杉矶公共空间的缺乏是一种弥补。

内部设计体现了这些隐密世界的精神。办公室、图书馆、社交场所和最重要的制作室都在不同的体积和空间中互相容纳和关联，并创造了一种平衡的张力。原有仓库的顶棚和空间依然完整，以一个恒久运动的概念延续了空间的整体流动性。

建筑内部的其中两边是大小相等的服务带。开放的厨房和接待区在第三边，第四边被保持很干净，那里采光很好，经过者能够从那里看到中央空间中两个被周围配室所围绕的、线条优美的房间矗立于一个浅水池之上。在整个空间中都能够体会到它们摄人的存在，它们有着光滑的灰色外表，就像一对正在玩耍的海洋哺乳动物；它们笨重的身体停留在半空中，像是要逃离水池的羁绊。当如镜的水面倒映了它们的影像时，轻重倒置的对照更加鲜明。当人工制造的水汽像雾一样覆盖两个物体时，它们就

Jigsaw

Location of project: west Los Angeles, California
Client/Owner: Jon Hopp and Traci Meger
Total square footage: 6000 sq.ft

Project team: Lawrence Scarpa, principal -in-charge, Peter Borrego, Angle Brooks, Fredrik N ilsson, Silke Clemens, Katrin Terstegen.

Contractor: Minardos, Inc.
Photography: Marvin Rand

A film editor lives in a fictional world, represented by the computer screen reflected upon them. To completely dive into this world, distractions, like light reflections, need to be blocked out, and so, typically, a film editor's world revolves around a hermetically closed black box. What does that mean for an architect who is asked to design a stimulating workspace and create an environment that allows for both social interaction and provides a place of seclusion?

Confronted with this challenge when they were commissioned to design Jigsaw, a film editing company the architects transformed this rough 5,000 ft² bow-truss warehouse into a new and unexpectedly world. Located in an industrial part of West Los Angeles within a rather featureless neighborhood the project incorporates independent forms separating the building envelope from the interior space, reminiscent of urban structures, as if to compensate for Los Angeles' lack of public space.

The interior is designed in the spirit of these hidden worlds. The program - offices, library, socializing zones and, most importantly, the editing rooms – are accommodated within a variety of volumes and spaces that relate to each other, creating a balance in tension. The volumes do not touch the ceiling and the original warehouse space can be read in its entirety. The circulation zone between them extends throughout the entire space, creating a constant notion of movement.

The perimeter of two sides of the interior is a belt of equally sized service spaces. An open kitchen and the reception area are on the third side, while the fourth side is kept clear, allowing daylight to enter through the windows and passers-by to get a glimpse of the central space. Taking up the entire stage that is surrounded by the ancillary rooms, two curvaceous volumes are suspended over a shallow pool of water. Their overwhelming presence can be sensed throughout the space. With a skin of sleek gray lead, they resemble a pair of playing ocean mammals; their heavy bodies in mid-air, escaping the pool, if not the building. The paradox of the inversion of heavy & light is reinforced as the mirror-like surface of the water below

完全被从地面分隔开。

进入空间时,两个物体被首先看到的部分是他们的尾端截面;两个白色的、怪异的屏幕对来访者而言就像两只巨大的眼睛。虽然屏幕给人的感觉是透明的,但人们还是无法从模糊的表面看透到另一侧。这也就引出了探究其中蕴涵的念想,这是空间中最有活力的元素,同时也是远离尘嚣的区域——编辑和制作师的工作间就在这里面。

屏幕就是采光的窗户,它创造了黑暗和光明之间一种模糊的感觉,这种光线有利于在电脑屏幕前的工作。窗户就是私人与公共区域的分界点,既提供了视觉的接触,同时又保留了空间的私密性。这样一来,长时间与世隔绝的电脑工作也显得与外界有所联系了。只有当你走得很近去观察这些轻巧、闪亮和模糊的物质时,你才会发现两扇屏幕不过是由普通的材料制成。一个从下到上被填满了乒乓球,另一个则被装满了亚克力珠子。从远距离来看,肉眼并不能发现其中的奥妙,而只是根据光线的方向,把这些元素或者缝隙看成是整个图画的一部分。因此,当光线落到屏幕上的时候,它们的立体感非常强。当眼睛看到填充之间的缝隙时,这个视觉定律又在工作室内部体现了相反的效果。当你走近时,一切都变得很清晰,就像一个阿拉伯屏幕,而里面的观察者什么也看不见。

物体和空间的关系也显现于整个空间中。建筑内部的物体拥有自己特别的形式,围绕它们的空间只是剩余的负空间。这些形成空隙的负空间为来访者、客户提供了等待区,在水池上形成了能够从一个不同角度欣赏美景的内部平台。

设计中变化多样的非正式空间为客户和员工提供了

reproduces their image and even more so, when occasionally artificially produced steam sheathes the two volumes like fog, completely detaching them from the ground.

The first thing that is visible of these volumes upon entering the space is their cut-off end pieces; two white, pixilated screens that confront each visitor like two giant eyes. Although the substance of the screens conveys the impression of transparency, one can not look beyond their fuzzy surface. This gives a clue of what is contained within. Despite being the most dynamic elements within the space, the most secluded and quiet areas – editing and producers' rooms – are placed here.

The screens are windows that filter the light from the outside, creating a fuzzy condition between darkness and light and thus enabling work on a computer screen. They act as an interface between the private and the public zones, providing visual contact and simultaneously guaranteeing privacy. This way the often-isolated work on the computer is soothed by contact to the outside world. From a distance, it is hard to tell what these screens are made of. Only at a closer look does this ethereal, glittering and out-of-focus substance reveal itself as quite ordinary materials: one window is filled bottom-to-top with ping-pong balls while the other is filled with acrylic beads. From a distance the eye doesn't read the details and instead connect either

the elements or the gaps to a whole picture, depending on the direction of the light. Consequently, they appear solid from the outside, where the light projects onto the screen. This visual principle is inverted on the interior of the studio as the eye reads the gaps between the filling. As one comes closer, what is beyond becomes sharper and the gaps more apparent, much like an Arabic screen, where the observer inside remains unseen.

This relationship between object and space is also discovered at a larger scale of the overall space. While the volumes within the building envelope have their own distinct form, the space around them is merely an in-between, residual space that takes on the negative shape of the volumes. These interstitial spaces form niches for informal encounters, waiting zones for clients, and interior terraces on the water, that offer views from a different vantage point.

The variety of informal spaces incorporated in the design allows clients and staff to have spaces to relax and to socialize. In addition, the entire entrance zone acts as a cafe that facilitates informal meetings and client interaction. Social activity is thrown into the limelight. On the opposite side of the cafe is the reception area, growing out of the linoleum floor material. The reception desk is deliberately moved from the entrance door, encouraging the visitor to absorb the space freely

Right: interior detail of Jigsaw

Jigsaw 室内细部

immediately.

The design of Jigsaw attempts to create a series of balanced tensions – between isolation and interaction, movement and static, light and heavy and between light and dark, generating a complex spatial experience, turning an office space into an inspiring

放松、社交的场所。另外，入口处的咖啡厅也是进行非正式会议和与客户交流的好地方，社交活动在此处的聚光灯下进行。在咖啡厅的对面是接待区，地上铺着油地毡，接待桌是特意从入口移过来的，这样做的目的是为了鼓励来访者更亲密地、自由地感受整个空间。

Jigsaw 的设计在隐居和交流、运动和静止、轻巧与沉重、光明与黑暗之间创造出一系列平衡的张力，形成了一种空间体验，并把办公空间变成了一个令人振奋的运动场。

Above, center, below:
renderings of Jigsaw, the
drawing is section plan

Jigsaw 效果图和平面图

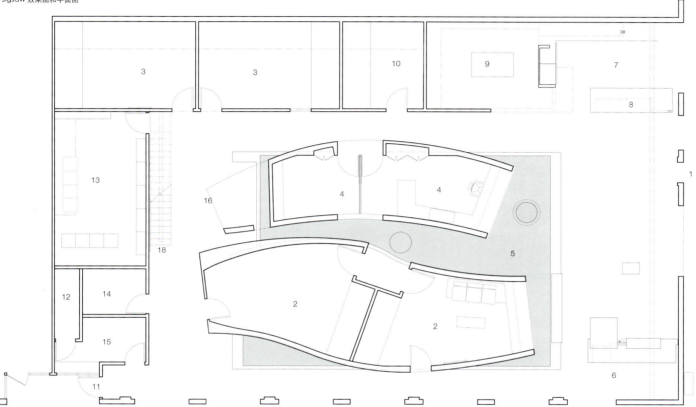

1. entry
2. edit studio
3. editing bay
4. office
5. reflecting pool
6. reception
7. kitchen
8. kitchen island/bar
9. conference room
10. music studio
11. service entry
12. mechanical
13. library/server
14. men's bath
15. women's bath
16. client waiting/group meeting
17. patio garden
18. mezzanine loft above

Below: interior view of Jigsaw
Jigsaw 室内

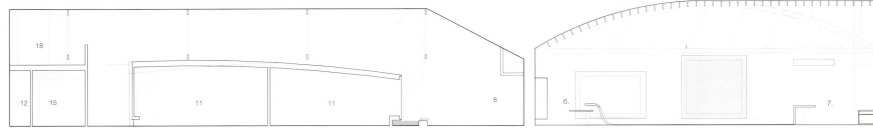

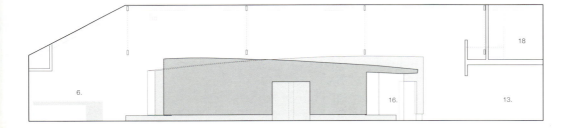

Above, below and opposite: house detail and house section plan, interior view of Jigsaw

建筑细部、剖面图和室内照片

1. entry
2. edit studio
3. editing bay
4. office
5. reflecting pool
6. reception
7. kitchen
8. kitchen island/bar
9. conference room
10. music studio
11. service entry
12. mechanical
13. library/server
14. men's bath
15. women's bath
16. client waiting/group meeting
17. patio garden
18. Mezzanine loft

Contents

Project Survey

Introduction with Article Review, Page 7. Colorado Court Page 23. COop Editorial Page 41. Nascent Terrain Page 59. Diva Page 67. Orange Grove Page 77. North Point Page 85. Jigsaw Page 95. **Solar Umbrella** Page 109. SM College Student Service Center Page 117. Bergamot Loft Page 125. Vail Grant House Page 139. The Firm Page 149. XAP Page 159. Reactor Films Page 171. Chronology Page 181 & Bibliography

Solar Umbrella

太阳能伞住宅
威尼斯，加利福尼亚州

在威尼斯的这栋住宅勇敢地开创了一个加利福尼亚州现代主义建筑的新纪元。建筑位于一块41英尺宽、100英尺长的地基上，原有的850平方英尺的住宅被改建成一个2 200平方英尺的新世纪住宅。

太阳能伞项目的灵感来源于1953年保罗·鲁道夫的"伞房"（Umbrella House），鲁道夫的伞房是太阳能屋顶的现代革新———一个在强烈光照气候中提供热防护的策略。此住宅的夫妇和他们的孩子参与到设计中，共同努力通过自身实践来实现建筑的可持续性。建筑师仔细考虑了整个地形的有利条件，并寻求可持续性居住的良机。活性的太阳能设计策略勾画出了住宅100%的能源独立机制，其间可回收的、可更新的和高性能的材料和产品随处可见。景观设置和风景的交融都考虑了它们的审美和实际影响力。这所住宅体现出的这些优雅的策略和材料，开拓了感性和演绎的潜力，获得了丰富有趣的感官和审美体验。

依据现有地形的有利条件，原有住宅的方向被转了180°。老建筑的正面和入口成为新建筑的后部，新建筑被调整为南向。这个举措使得建筑更加完美并沐浴在南面温暖的阳光中。建筑的南立面被太阳能板包裹起来，形成的屋顶定义了这所住宅的形式表现语言。太阳能屋顶成为建筑的保护屏障，使之不受南加州强烈的光照下产生的热量的影响。太阳能板并非反射阳光，而是将其丰富的资源转化成为可用的能源，为整个住宅提供了100%电力。就像此住宅的许多设计一样，这个太阳能屋顶拥有多元而且丰富的意义，在功能、形式和效果体验方面显示了多种作用。

Solar Umbrella

Location of Project: Venice, California
Client/Owner: Angela Brooks and Lawrence Scarpa
Total Square Footage: 1,100 sq. ft.(new) 780 sq. ft.(remodeled)

Project team: Lawrence Scarpa and Angela Brooks - Principals- in-Charge. Peter Borrego, Angela Brooks, Anne Burke, Vanessa Hardy, Ching Luk, Gwynne Pugh, Lawrence Scarpa.

General Contractor: Lawrence Scarpa and Angela Brooks
Photography: Marvin Rand

Nestled amidst a neighborhood of single story bungalows in Venice, California, the boldly establishes a precedent for the next generation of California modernist architecture. Located on a 41' wide x 100'-0" long through lot, the addition transforms the architects' existing 850 square foot bungalow into a 2200 square foot residence equipped for responsible living in the twenty-first century.

Inspired by Paul Rudolph's Umbrella House of 1953, the Solar Umbrella provides a contemporary reinvention of the solar canopy — a strategy that provides thermal protection in climates with intense exposures. In establishing the program for their residence, which accommodates the couple and their one child chose to integrate into the design, principles of sustainability that they strive to achieve in their own practice. The architects carefully considered the entire site, taking advantage of as many opportunities for sustainable living as possible. Passive and active solar design strategies render the residence 100% independent from the grid. Recycled, renewable, and high performance materials and products are specified throughout. Hardscape and landscape treatments are considered for their aesthetic and actual impact on the land. The Residence elegantly crafts each of these strategies and materials, exploiting the potential for performance and sensibility while achieving a rich and interesting sensory and aesthetic experience.

Taking advantage of the unusual through lot site condition, the addition shifts the residence 180 degrees from its original orientation. What was formerly the front and main entry at the north becomes the back as the new design reorganizes the residence towards the south. This move allows the architects to create a more gracious introduction to their residence and optimizes exposure to energy rich southern sunlight. A bold display of solar panels wrapping around the south elevation and roof becomes the defining formal expression of the residence. Conceived as a solar canopy, these panels protect the body of the building from thermal heat gain by screening large portions of the structure from direct exposure to the intense southern California sun. Rather than deflecting sunlight, this state of the art solar skin absorbs and transforms this rich resource into usable energy, providing the residence with 100% of its electricity. Like many design features at the house, the solar canopy is multivalent and rich with meaning—performing several roles for both functional, formal and experiential effect.

By removing only one wall at the south, the architects maintain the primary layout of the existing residence. The original bungalow, which was tightly packed with program (kitchen, dining, living, two bedrooms and a bath) is joined by a sizable addition to the south which includes a new entry, living area, master suite accommodations, and utility room for laundry and storage. The kitchen, which once formed the back edge of the residence, opens into a large living area, which in turn, opens out to a spacious front yard. An operable wall of glass at the living area delicately defines the edge between interior and exterior. An unbroken visual corridor is established from one end of the property to the other. Taking cues from the California modernist tradition, the architects conceive of exterior spaces as outdoor rooms. By creating strong visual and physical links between outside and inside, these outdoor rooms interlock with interior spaces, blurring the boundary and creating a more dynamic relationship

Below: rendering of Solar Umbrella
太阳能伞住宅效果图

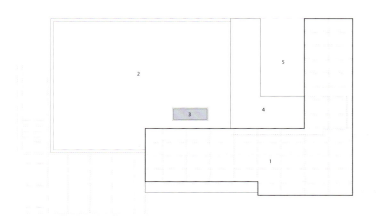
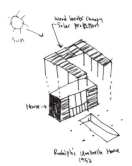
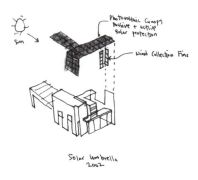

ROOF PLAN

1. PHOTOVOLTAIC ROOF PANEL
2. (E) ROOF
3. (E) SKYLIGHT
4. LOWER ROOF
5. PATIO BELOW

Opposite: roof plan and rendering of Solar Umbrella
屋顶平面及太阳能伞住宅效果图

Below: rendering and modeling of Solar Umbrella
太阳能伞住宅模型及效果图

between the two. The entry sequence along the western edge of the property further demonstrates this concept. A cast in place concrete pool provides a strong landscape element and defines the path to the front entry. Upon reaching the entry, the pool cascades into a lower tier of water that penetrates and interlocks with the geometry and form of the residence. In a move that reinvents the welcome mat, stepping stones immersed in the water create an initiatory rite of passage into the residence as the visitor is invited walk across water. The distinction between outside and inside is once again blurred.

The master suite on the second level reiterates the strategy of interlocking space. Located directly above the new living area, up a set of floating, folded plate steel stairs, the bedroom strategically opens onto a deep covered patio which overlooks the garden. Conceptually reminiscent of R.M. Schindler's Kings Road Residence, this patio extends the bedroom area outdoors, creating the sensation of a sleeping loft exposed to the exterior. This deep porch carves out an exterior space within the visual bounds of the building envelope and provides the front elevation with a distinctive character. What appears to be a significant area of the second floor is actually never enclosed but rather, it is protected by the planes which wrap around it.

Below: rendering of Solar Umbrella
太阳能伞住宅效果图

Opposite: Site view and floor plan
地形环境和楼层平面图

建筑的原平面结构基本被保留，仅仅拆除了南面的一面墙。原来的平房设计得比较紧凑（包括厨房、起居室、两个卧室和一个浴室），与南边的一个大型的附加部分相连。该附加部分包括一个新的入口、客厅、主套房、洗衣房和储存间。厨房位于建筑的后部，通向一个大的客厅，穿过客厅则是一个很大的前庭。客厅的可操作玻璃墙分隔开了建筑的内部和外部空间，其间一条连续的走廊贯穿了整个建筑。设计借鉴了加利福尼亚现代主义建筑的特色，将外部空间构思成户外场所。建筑

A dynamic composition of interlocking solid and void creates a richly layered depth to the design. Transparency through the house allows views to penetrate from front to back. The structure appears to sit lightly upon the land. Formal elements along these visual corridors—i.e. stairs, bearing walls, structural columns, guardrails, built-in furniture and cabinetry— vary in density, color and texture. Light penetrates the interior of the residence at several locations. A series of stepped roofs, glazed walls, and clerestory windows broadcast light from multiple directions. Light and shadow—ephemeral and constantly changing effects—become palpable formal tools that enliven the more permanent and fixed elements of the design. Together, all of these components establish an effectively layered composition rich in visual and formal interest.

Throughout the residence, the architects resourcefully take materials and contextually reposition them as design elements. Solar panels, conventionally relegated to a one dimensional utilitarian application, define envelope, provide shelter and establish a distinctive architectural expression. Homosote, an acoustical panel made from recycled newspaper is palm-sanded and used as a finish material for custom cabinets. OSB (oriented strand board) a structural grade building material composed of leftover wood chips compressed together with high strength adhesive, becomes the primary flooring material where concrete is not used. Sanded, stained and sealed, the OSB floor paneling provides a cost effective and materially responsible alternative to hardwood. Materials are selected for both performance and aesthetic value. Metal stud construction replaces conventional wood framing. Recycled steel panels, solar powered in-floor radiant heating, high efficiency appliances and fixtures, and low v.o.c. paint replace less efficient materials. Decomposed granite and gravel hardscape are used in place of concrete or stone. Unlike their impervious alternatives, these materials allow the ground to absorb water and in turn, mitigate urban run-off to the ocean. Drought tolerant xeriscaping compliments the textures and palette of the building while providing a low maintenance, aesthetically appealing landscape.

内外的视觉和物理结合使户外场所和内部空间融合在一起，二者之间的界限模糊，形成了一种更加动态的关系。建筑西侧的系列入口也同样展示了这个概念。位于正面入口通道的水池是该建筑的重要景观要素，在到达入口后，水池中的水流会降低一层，并和住宅的几何形态联系在一起。建筑新设计了迎宾毯和浸入水中的石阶，来访者可由这条水中小道进入住宅，这使得建筑的内外空间关系再一次被升华了。

第二层的主套房再次诠释了将所有空间融为一体的策略。一组悬浮的、层叠的钢制阶梯位于新客厅的上方，卧室通向一个可俯瞰花园的加顶内院。这个内院从概念上再现了R.M申德勒设计的国王路住宅（R.M Schindler's Kings Road Residence）的特点。内院将卧室区延伸到了户外，创造了一个感觉上的露天眠憩阁楼。这个深层的门廊在建筑的内部视觉范围内勾画了一个外部空间，创造了风格独特的建筑前立面。第二层的重要特点是完全敞开的形式，但是被四周围绕的太阳能板所保护。

一个动态的互连的固体和空间为整个设计创造了丰富的层次和深度。建筑的透明度使人们的视觉可以从前面贯穿到后面。 整个结构轻巧地立于土地之上。空间视觉长廊中的那些形式元素——比如楼梯，支撑墙，结构柱，栏杆和粗、细工家具在密度、颜色和质地上都各不相同。一系列的台阶形屋顶、玻璃墙和天窗将射入住宅内不同地点的美丽阳光向四面八方折射，光和影、短暂的和不断变幻的效果成为使设计中固定元素更生动的、具体可见的形式工具。总之，所有这些成分共同建立了视觉、形式丰富的多层复合体。

综观整个住宅，建筑师广泛地采用了各类材料，从不同的角度来重新定义这些设计元素。太阳能板在传统意义上只是一个应用工具，而现在，它具有了遮蔽和表达独特建筑风格的功能。一种用回收报纸制成的纸板——Homosote在手工打磨后被用作于橱柜的表面材料。OSB（定向刨花板）是一种结构性建筑材料，内含经过强力压缩的碎木屑，因拥有优良的硬木性能和视觉审美价值而被用于地面。金属螺栓结构取代了传统的木结构，低效材料被可回收钢板、太阳能供热系统等高效材料和设备以及低v.o.c油漆所代替。可分解的、透水性能良好的花岗石和卵石替代了混凝土，这些材料能够促进地面吸水，从而减少城市废水流向大海。耐旱的xeriscaping材料优化了建筑的质感和颜色，这种极具美感的材料只需要低廉的维修费用。

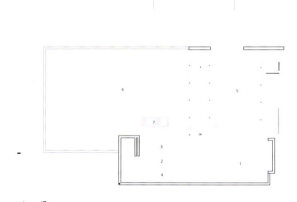

SECOND FLOOR PLAN

1. MASTER BEDROOM
2. MASTER BATHROOM
3. SHOWER/TUB
4. CLOSET
5. PATIO
6. (E) ROOF
7. (E) SKYLIGHT

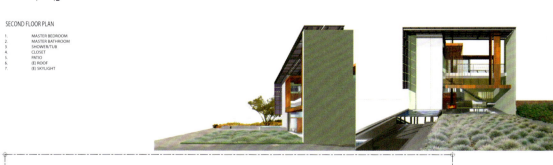

FIRST FLOOR PLAN

1. LIVING
2. (E) DINING
3. (E) KITCHEN
4. (E) BEDROOM
5. (E) STUDY
6. (E) BATHROOM
7. (E) CLOSET
8. WATER POND
9. BAMBOO PLANTER
10. BATH
11. LAUNDRY
12. A.V. CENTER

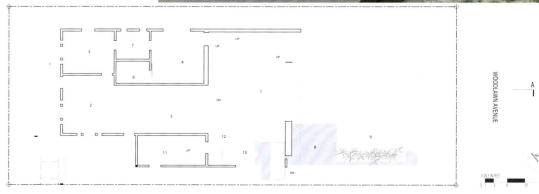

Contents

Project Survey

Introduction with Article Review, Page 7.
Colorado Court Page 23. COop Editorial Page 41. Nascent Terrain Page 59. Diva Page 67. Orange Grove Page 77. North Point Page 85. Jigsaw Page 95. Solar Umbrella Page 109. **SM College Student Service Center** Page 117. Bergamot Loft Page 125. Vail Grant House Page 139. The Firm Page 149. XAP Page 159. Reactor Films Page 171. Chronology Page 181 & Bibliography

SM College Student Service Center

圣莫尼卡学院学生服务中心
圣莫尼卡,加利福尼亚州

该项目原址是占地4 000平方英尺的校园书店，位于圣莫尼卡学院中心并沿着校园的中轴线紧邻校园食堂和学生服务中心。经扩建和改建后的新设施将成为校园未来的多功能中心。

扩建后的校园书店增加了一个新的入口和存包处，学生们可以在中心活动时存下他们的书包，在离开时再从中心外面的窗口把包取走。

学生服务中心这个19英尺高的增扩建筑完全由可回收金属和环保材料建成。高顶棚和附加的功能设计为建筑提供了充足的自然光和空气流通，以减少对空调、暖气和其他机械系统的需求。这座建筑是校园中惟一一个没有机械供热和制冷系统的节能和可持续性建筑，家具和木制设施由一种再生木板(OSB)制成。

建筑的入口更像是一个学生们能够聚会和结识新朋友的走廊。屋顶转变成为一个垂直的臂状物，一直达到地面并成为遮挡西晒的门廊顶棚。这个"手臂"以一个简单的支点轻轻地接触着地面，学生们下课后，可以像走过校园大道一样在建筑内自由地穿行。

这个新的学生服务中心暂时还是校园的书店。在新的书店落成之后，学生服务中心将被作为进行各类学校活动和聚会的多功能设施。

这个项目采用了一个复杂的阶段性计划使得原有的书店在施工中可以继续使用。

SM College Student Service Center

Location of Project: Santa Monica, CA
Client/Owner: Santa Monica College
Total Square Footage: 3,500 sq. ft.

Project team: Lawrence Scarpa, Angela Brooks, Silke Clemens, Michael Hannah, Vanessa Hardy, Anne Marie Burke, Ching Luk, Fredrik Nilsson, Tim Petersen, Gwynne Pugh, Katrin Terstegen.
Engineering: Vohn Martin Assoc.
General Contractor: TriMax Inc.
Photography: Marvin Rand

This new expansion and remodel to the existing 4,000 sq. ft. campus bookstore is located in the center of the Santa Monica College campus, adjacent to the campus food court and central student services along the main axis across the campus. This new facility is designed to be the future hub on the campus.

The expansion provides the existing campus Bookstore with a new entrance, complete with a backpack drop-off counter, where students check their packs and bags while in the Center and retrieve them at a staffed window from outside the building once they leave.

The Student Services Center expansion is a single story; 19' high building made with entirely recycled metal and environmentally friendly materials. The high ceiling and additional design features provide significant natural lighting and ventilation eliminating the need for air conditioning, heating and other mechanical environmental systems. The building is the only energy efficient and sustainable structure on the entire campus that is without mechanical heating and cooling. Cabinetry and woodwork are constructed of Oriented Strand Board (OSB), a recycled wood product.

The entry of the building acts much like a porch where students can congregate in the shade and have impromptu encounters with each other. The roof transforms into a vertical arm that reaches to the ground and screens the porch from the western sun. The "arm" touches the ground lightly at a single point so that students can pass thru the building as they travel from class to class across the campus' main thoroughfare.

The new Student Services Center will act temporarily as the campus bookstore annex until the new bookstore is completed. Upon completion of the new bookstore the Student Services building will be converted into the campus Conference Center. It will also be used as a multi-purpose facility for a variety of campus wide special events and meetings.

The project involved an intricate phasing plan to enable the continuous operation of the existing bookstore during construction.

Below: the entry of the
Student Service Center

学生服务中心入口

Opposite: exterior view,
plan and section drawings

建筑外观，平面和剖面图

SM College Student Service Center

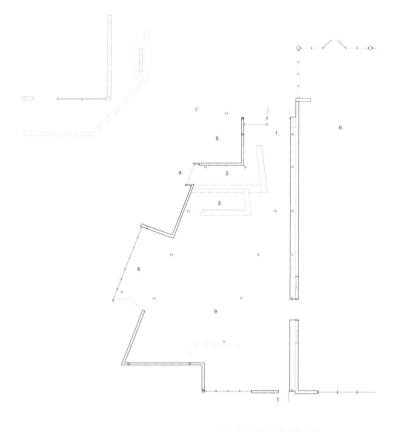

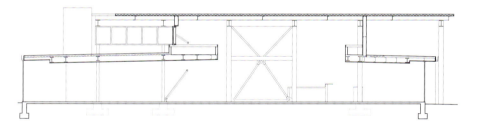

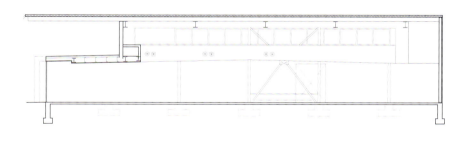

1. ENTRY VESTIBULE
2. BOOK CHECK-IN
3. INFORMATION
4. BOOK PICK-UP
5. PORCH
6. EXISTING BOOKSTORE
7. EMERGENCY EGRESS
8. DISPLAY WINDOW
9. MERCHANDISE AREA

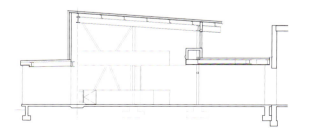

SM College Student Service Center

Contents

Project Survey

Introduction with Article Review, Page 7.
Colorado Court Page 23. COop Editorial Page 41. Nascent Terrain Page 59. Diva Page 67. Orange Grove Page 77. North Point Page 85. Jigsaw Page 95. Solar Umbrella Page 109. SM College Student Service Center Page 117. **Bergamot Loft** Page 125. Vail Grant House Page 139. The Firm Page 149. XAP Page 159. Reactor Films Page 171. Chronology Page 181 & Bibliography

Bergamot Loft

伯格莫特艺术家阁楼
圣莫尼卡，加利福尼亚州

项目位于伯格莫特站点（Bergamot Station）——一个国际知名的艺术中心；其中包括45栋改建成艺术博物馆的工业建筑和最近完工的圣莫尼卡艺术博物馆。这个项目包括一个底层工作室及画廊和上层3个艺术家生活及工作的阁楼空间。整个项目于1999年竣工，每平方英尺建筑预算87美元。

这个项目最大的挑战在于：如何在保留伯格莫特站点原有仓库建筑形态的连续性和紧密性的同时实现形式和材料的实验和革新。所以，该项目充分考虑了在原有建筑基础上的材料革新：依据周围环境和现有环境中的材料形态，利用冷轧钢和透明莱克桑（Lexan）板为建筑在细节上传达了不同的诠释，给予它独树一帜的特性。建筑充分利用了工业景观的地形特点并置身于其中，新建筑的正面随着波纹状金属贴板延伸至由周边建筑形成的居住空间中。建筑正面的窗户和表面材料的层叠起伏造就了整个建筑的生动活泼，其形式和材料的多样化为未定义的余留空间带来无限动感，使得其成为建筑自身和周遭建筑的庭院广场。事实上，余留的空间现在经常被用来举办户外接待和特别活动。

Bergamot Loft

Location of Project: 2415 Michigan Ave, Santa Monica, California, USA
Client/Owner: Bergamot Limited Partners, Inc., Wayne Blank
Total Square Footage: 14,000 sq. ft.

Project team: Lawrence Scarpa - Principal- in-Charge. Jackson Butler, Peter Borrego, Angela Brooks, Anne Burke, Anne Marie Kaufman Brunner, Tim Peterson, Gwynne Pugh, Lawrence Scarpa.

Furniture, Carpets and Fixture Design: Peter Borrego, Jackson Butler and Lawrence Scarpa.

Structural Engineering: Gwynne Pugh and Joe Castorena.
General Contractor: Pegan Construction
Photography: Marvin Rand and Benny Chan

Program: This project is located at Bergamot Station, an internationally known art center comprised of a series of industrial buildings converted into 45 art galleries including the recently opened Santa Monica Museum of Art. The program includes a ground level studio/gallery space with three artist live/work loft spaces above. The project completed construction in 1999 and was realized with a construction budget of $87/sf.

The fundamental challenge in this project was determining how to maintain continuity and coherence with the character of the existing industrial warehouse buildings at Bergamot Station without compromising formal and material experimentation and innovation. Thus, this project evolved as a carefully considered response to its context: a primary palette of materials was established with regard to the existing industrial materials at the site. Corrugated metal, steel and glass blend in with the surrounding context while cold rolled steel and translucent lexan panels create moments of distinction in the details of the building that set it apart and help establish its idiosyncratic identity. The building takes advantage of its unique siting amongst the industrial landscape. Nestled in between existing warehouse buildings on a narrow site, the facade facing the interior of the site unfolds itself gracefully along a canted corrugated metal plane that extends itself into the residual space produced by the adjacent buildings. The facade is further animated by window boxes and planes articulated as volumes unto themselves, which push and pull from the facades primary folds. The formal geometries and material richness of this facade have a dynamic effect on the leftover space-turning what was once experienced as in between and perhaps undefined space into one that now flourishes as a kind of courtyard or piazza for itself and the surrounding buildings. In fact, this residual space

Below: exterior view of the Bergamot Loft

伯格莫特阁楼外观

Bergamot Loft

Opposite: interior view of the Bergamot Loft, entry and stair

伯格莫特阁楼室内，入口和楼梯

is now often used to host outdoor receptions and special events. The building's south facing facade which fronts the public street provides a personality foil for its more geometrically complex and dynamic counterpart on the Arts Complex facing facade. More reserved yet still sculpturally articulate; the formal resolution of this elevation is calmer and more grounded. A flat plane of corrugated metal is broken by pristine rectangular volumes of lexan, concrete block, cold rolled steel and glass which recede to varying degrees creating a constrained yet elegant relief and textural complexity. The exterior of this building never strays to far from its industrial origins and therefore maintains a respectful coherence with its context. Yet, the architects still capitalize on each contextual and formal opportunity in order to create distinction and enhancement for the site and its surrounding context.

The interior of the building is simply organized. The ground floor features an open plan that allows maximum flexibility of use and reuse. Simply treated in its finishes and details, the ground floor interior is conceived as a vessel for program to animate. A separate entrance leads to the three artist loft units above. Each of these maximizes its potential for spaciousness and light while also creating moments of intimacy and enclosure. Each unit has a fluidity of space and circulation that creates a sensation of open airiness even though they are flanked between buildings on either side and only feature minimal windows to the outside from within the units. Each interior is deliberately treated as a simple volume or shell in which the distinct elements of the space can more clearly emerge. Polished concrete floors create a uniform field condition at the ground plane. Plainly painted drywall walls and an exposed steel truss and metal deck roof system continue the effect of creating a quiet background field in which feature elements can construct spatial and textural complexity.

Each unit is split level. Public spaces are on the primary level while a bedroom space occupies a loft, which overlooks the public living space. By organizing this split-level condition, the architects could create a double height space, which adds to the feeling of vastness in each unit. While occupying the most private, intimate corner of the unit, each bedroom area is awash in light from a continuous band of skylights along one edge of the building. The double height space on the level below is left uncluttered and open while bathroom; kitchen and circulation cores are tucked into the space below the loft. The space becomes most animated in its details, which distinctly emerge from the more tranquil background. Two flights of wood and steel stairs stand across from each other. Sculptural presences-one leads to a balcony space, the other

面对公共大街的建筑南面墙以其个性化的表面设计体现了更加复杂的几何特点和动感对照。设计上凸显的雕塑感使建筑显得稳重大方。建筑外部平淡的波纹状金属平面被质朴的莱克桑矩形物、水泥砖、冷轧钢和玻璃打破，创造了一个优雅的、充满质感的浮雕复合体。建筑外部并没有过于脱离工业本质，因而与周围的环境浑然一体。当然，建筑师们依然不会放过任何在环境和形式中创造独特效果的机会。

建筑内部的结构十分简约。开放式的底层有利于最大限度的空间使用和再利用，其修饰和细节上的设计也很简单，其构思就是成为激活整个建筑的载体。一个独立的入口通向上层的3个艺术家的阁楼单元。每一个单元都在展现其内在私密性的同时尽可能地扩展空间和接受阳光。虽然三个单元都处于建筑的两侧并且朝外的窗户比较小，但是其良好的空间流动性保证了空气的流通。每一个室内空间都很独特而且层次分明，磨光的水泥地板对整体风格起到了统一协调的作用。平坦粉刷过的墙面、裸露的钢架房梁和金属顶棚等元素构成了一个宁静的结构空间和质感的复合背景效果。

每个单元都是分层式的。阁楼是卧室，可俯瞰第一层的公共生活区。在设计这个分层环境时，建筑师以2倍加高的空间来扩大了每个单元的空间感。占据着单元最隐私角落的卧室通过建筑旁边的一系列天窗获取足够的采光。这个双倍空间的下方非常整洁，浴室、厨房和流通中心都聚集在这里。整个空间的各个细节都活泼生动，在宁静的背景中显得尤其别具一格。两段木制和钢制楼梯相互交错，其中一个通向阳台，另一个通向卧室。并列的两者展现了一种诗一般的平衡感：一个稍大的占主要位置，另一个较小但很坚固。 建筑师在

Opposite:interior view of
the Bergamot Loft
伯格莫特阁楼室内

to the sleeping lofts. When viewed in juxtaposition, they hold each other in poetic balance, one larger and dominant, the other more petite yet solid. In another unit, a poured in place concrete fireplace acts as an anchor to the open plan, which unfolds around it.

Consistent throughout this project is the desire to create a space that capitalizes on its ability to create an elegant and refreshing fluidity and coherence appropriate to its context while also inserting elements of distinction and complexity that help create this building's unique personality.

另一单元中设计了一个美观的水泥壁炉，它成为了周围开放空间的中心。

在这个项目中始终贯彻的思想就是创造这样一个空间——它能够以优美和新鲜的动感充分融入其环境，并同时以其独特的元素和复杂性来塑造建筑与众不同的特性。

Above: dusk exterior view of the Bergamot Loft
伯格莫特阁楼黄昏外景
Opposite: interior view
室内

Bergamot Loft

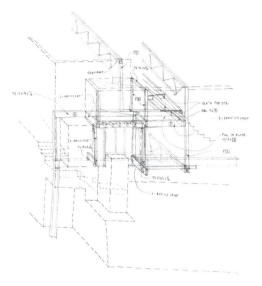
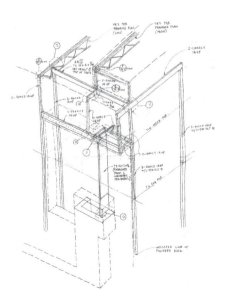

Opposite: exterior view of the Bergamot Loft
伯格莫特阁楼外观

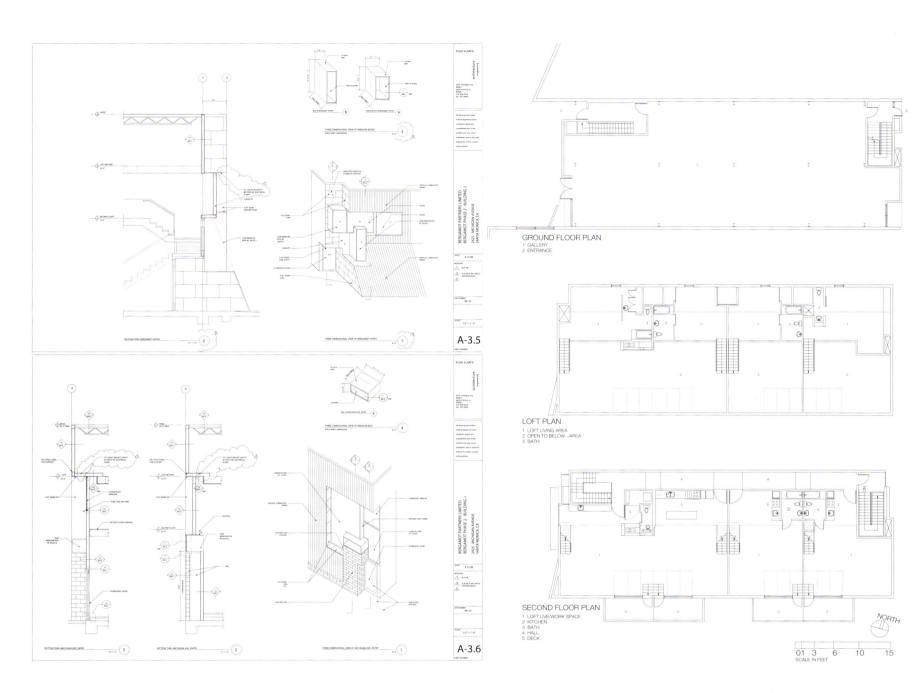

Bergamot Loft

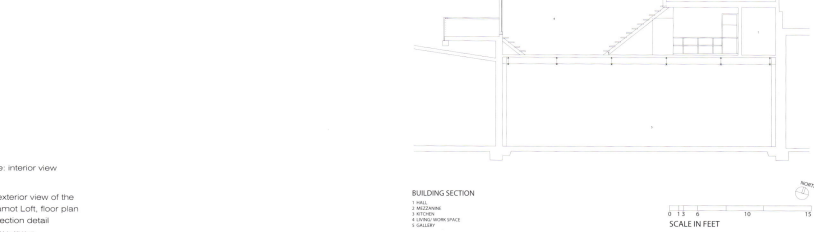

Above: interior view
室内

Left: exterior view of the Bergamot Loft, floor plan and section detail
伯格莫特阁楼外观
楼层平面和剖面细部图

BUILDING SECTION
1 HALL
2 MEZZANINE
3 KITCHEN
4 LIVING/ WORK SPACE
5 GALLERY

Contents

Project Survey

Introduction with Article Review. Page 7.
Colorado Court Page 23. COop Editorial Page 49. Nascent Terrain Page 59. Diva Page 67. Orange Grove Page 77. North Point Page 85. Jigsaw Page 95. Solar Umbrella Page 109. SM College Student Service Center Page 117. Bergamot Loft Page 125.

Vail Grant House Page 139.

The Firm Page 149. XAP Page 169. Reactor Films Page 171. Chronology Page 181 & Bibliography.

Opposite: drawing and modeling of the Vail Grant House

韦尔住宅模型和住宅地形图

Below: Vail Grant House in the site view

韦尔住宅地形环境图

Vail Grant House

ZONING ENVELOPE

Vail Grant House

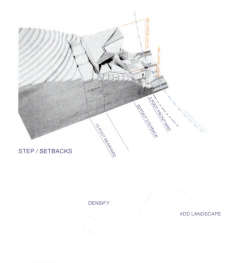

STEP / SETBACKS

DENSIFY   ADD LANDSCAPE

PROGRAM

Vail Grant House

Location of Project: 1820 Silverwood Terrace, Silverlake, CA
Total Square Footage: 1,800 sq. ft.

Project team: Lawrence Scarpa, Angela Brooks, Silke Clemens, Michael Hannah, Vanessa Hardy, Ching Luk, Fredrik Nilsson, Tim Petersen, Gwynne Pugh, Katrin Terstegen .

Engineering: Luis Vasquez
Photography: Marvin Rand

A topography sculpted of folded, skewed metal planes, the Vail House seems to enter into a love affair with the hill, blurring the boundaries between the natural and the artificial.

The design of the Vail House was generated by the integration of two disparate forces: the mundane requirements of the regulations imposed by zoning codes, economic constraints and the technical challenge of building on a steep hillside, and on the other hand the careful attention to the very specific condition of the site itself and to its surroundings. This made the project a unique expression of the generic and the specific.

The property is located in Silverlake adjacent to a Neutra house. An architectonically rich neighborhood in Los Angeles emblematic for the city's Modern movement, Silverlake represents a typical residential area in Los Angeles, overlaying a densely knit urban fabric with a layer of private outdoor spaces.

The clients, a young couple, wanted to see these characteristics carried on into their house while looking for an economical, environmentally friendly design.

As opposed to a classically Modern approach, where the site conditions and the landscape are perceived as a mere backdrop for the building and remain untouched, this project is in large part directly related to the topography and engages with the landscape, diving into the hill at points and breaking away from it at others. Consequently, the building becomes an abstracted, facetted reading of the landscape that contains it.

韦尔格兰特住宅
银湖，加利福尼亚州

韦尔格兰特住宅是一个嵌入自然的层叠倾斜面的建筑，它就像与层层青山陷入爱恋之中，不经意间将自然和人工的界限模糊化。

韦尔格兰特住宅的设计来自两种不同力量的结合：地区法规的要求与经济限制，斜坡建筑的技术性挑战；另外对该地点和周边地区的细心勘测使得整个项目无论从整体还是细节上都表达得淋漓尽致、独具匠心。

该建筑位于尽显城市现代发展的洛杉矶著名建筑——纽特拉住宅（Neatra House）旁边的银湖。银湖是洛杉矶典型的住宅区，它是拥有一层私人户外空间的高密度都市住宅区。

客户是一对年轻的夫妻，他们希望在造价经济和不影响周围自然环境的前提下把这些风格带入他们的房子中去。

有别于传统思路，该项目将地形条件和周围风景看作是建筑的重要背景元素而被原样保留。韦尔住宅被直接结合为地形景观的一部分，从一个坡点嵌入，并在坡面上凸显出来。因而，这座建筑在如画的风景中显得别具一格。

虽然这个建筑看上去与地形直接相关联，但是事实上这个造型很大程度是源于一些斜坡建筑政策法规。地区法规要求面向街道的建筑高度稍低，允许山上的建筑高一些。这样就意味着可以在靠近街道处修建房屋，在与周围的小规模建筑群建立联系的同时可以增加建筑的高度，这样就能充分利用地形

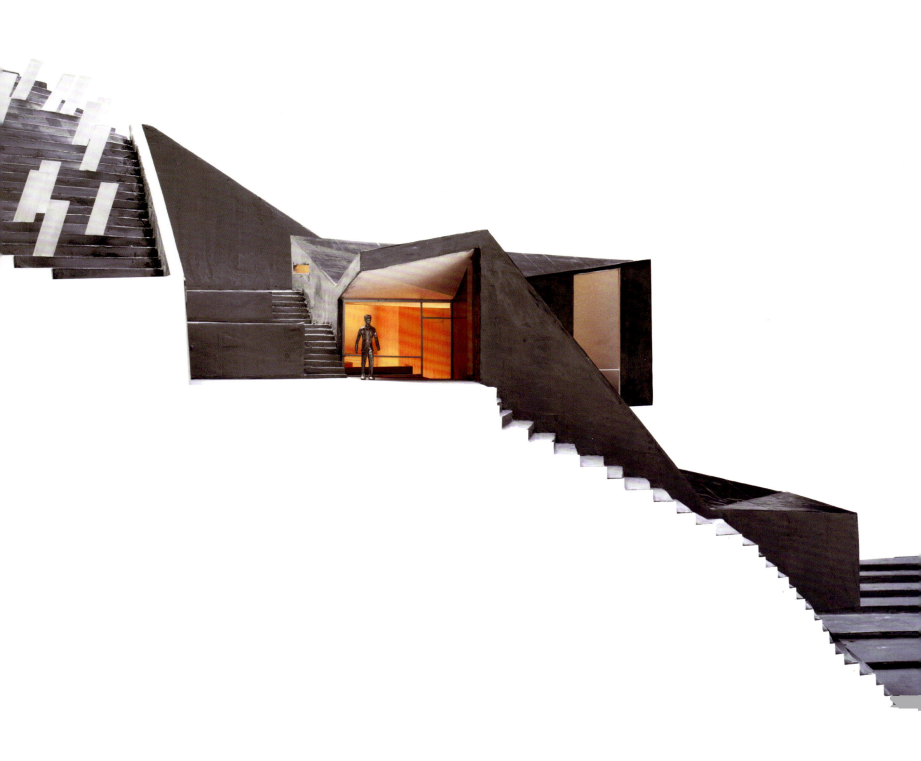

Below: renderings of Vail Grant House

韦尔住宅效果图

Although the building appears to be a direct response to the topography, much of its shape actually derives from a translation of the complex setback and stepback requirements of the hillside ordinances as they relate to this site. The zoning codes require a lower building height towards the street and permitted a taller structure further up the hill. By that means, it was possible to build relatively close to the street and establish a relationship to the smaller scale in the surroundings, while being able to increase the height further back in the lot and thus taking advantage of the spectacular views.

Organized internally through a succession of planes that follow directly the course of the topography, the movement through the building reflects the experience of walking up the hill.

The building volume is created by a simple extrusion of a square, a neutral elongated twisted box that is projected into the site and sculpted along its contours. The folded roof is skewed where directed views or openings are desired.

The building's movement on the site describes a spiral that begins at a lower point closest to the street, travels up the hill, and then turns back towards the street and the lake, overlooking itself and creating an enclosed court in the center.

This court serves as an entry to the building, covering a parking garage underneath, and is the first in the succession of planes. An 18' high entrance hall divides the building into its private and public domains and demarks its upper and lower part. From here, the building slopes down to the private realm – the children's room and the master bedroom, where the continuous, warping space ends in a window that takes up the entire section of the volume. The bathroom, steps up in the other direction, is completely dug into the hill and is lit by a skylight in the patio above. On the other side of the entrance hall, the stair leads up to the kitchen and dining room. A long window towards the hill allows air to circulate through the building from a low window in the living area and provides the necessary cross ventilation. A built-in stand-alone wooden box contains the guest bathroom and defines the transition between the kitchen and the living zone.

An excavation out of the corner of the building at the point where the roof is lifted makes room for a covered patio that can be entered through the dining area. From here, there is access to the back yard, a 15' wide excavated space that continues the succession of planes from the inside, creating a transition from the interior landscape to the hill.

On the opposite side of the kitchen/dining area, the volume folds away from the hill, moving downwards and again towards the street, when it is dramatically cut off, leaving a framed view overlooking the Silverlake reservoir.

The physical usage of the landscape was not as important as the containment of the building within the landscape, the creation of an artificial landscape inside and the experience of the distant views. The folding of the volume and the openings on its interior facade make it possible to inhabit the space and simultaneously to view it from within, across the void and to itself once more, creating a condition where the observer becomes the object of his own observation. The path of one's own and the building's movement can be retraced through the openings on the interior skin.

One of the central questions of this project was how to achieve an economical design on a site that was almost impossible to build upon. An early scheme, a serene, two-story building that maintained the same absolute height throughout and established no relationship to the landscape, proved to be uneconomical because it required huge retaining walls. It became necessary to develop a strategy that would keep the retaining walls as low as possible. This was accomplished in the final scheme by adjusting the building height to the topography, using Structural Concrete Insulated Panels (SCIP), a lightweight, easily to assemble system

获得广阔的视点。

整个建筑直接连续地沿着地形路线设计，穿过建筑就宛如体验登山的过程。

建筑空间是一个简约的正方形突出物，一个浅灰色、拉长并扭曲的箱体。它呈突出状，轮廓被清晰地勾勒出来。层叠的屋顶是倾斜的，透过那里能看到秀美的风光。

建筑从临街最低点呈螺旋状上升，穿过山坡，然后掉转回来面向街道和银湖，并在中心地段形成了一个封闭的庭院。

这个庭院是建筑的入口，庭院地下的停车场是这个连续建筑的第一层。18英尺高的入口大厅将整个建筑分为私人和公共部分，也区分了建筑的上下部分。从这里，建筑下斜向一个私人领域——这个空间被孩子们的房间和主卧室占据，尽头是一个

Above: renderings of Vail Grant House
韦尔住宅效果图

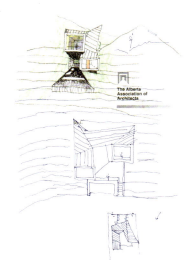

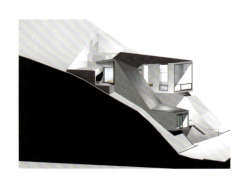

Left: rendering and model of Vail Grant House
韦尔住宅效果图和模型

大窗户。在上斜方向，浴室被完全建入坡体中，浴室光线来源于院子的天窗。在入口的另一边，有阶梯通向厨房和餐厅。新鲜空气可以从起居室的低窗进入，以保证建筑内的循环通风。一个内置的独立木制包厢内设有一个客人浴室，它是厨房和起居区的过渡。

在建筑的一角挖掘出一个带顶的院子，可以通过起居区进入。从这里还可以通向后院——一个开凿出的15英尺宽的空间，它是内部空间的延续，也是内部景观与外部青山的转换点。

厨房/饭厅的对面空间呈折叠状从山坡向下延伸，一直通向街道——那里是建筑的尽头，远眺则是银湖美丽的风光。

自然景观的运用并不像建筑置身其中所创建的内在人工景观那么重要。建筑空间的层叠和开放的内观使得空间被占据的同时，视觉能从内到外多次穿过空间。这使观察者成为他自己观察的对象。一个人的行走路线沿着建筑的走向在内部的通道中不断地迂回。

这个项目的中心问题是如何在一个几乎不可能实现经济造价的地形上完成设计。早期的计划是建造一个稳重的两层建筑，它的高度与坡面相同并且与周围的景观毫无关系。这个计划并不节省开支，因为这需要建巨大的支撑墙。因此，建造尽量低的支撑墙是必要的选择和最终的解决方案。设计师根据地形的特点对建筑的高度进行了调整，使用了轻便、容易安装、比水泥墙更经济实用的SCIP（结构隔缘墙板）。这样，在内部空间的基础上开发了一个外部的构架，而内含的轻型冷轧钢可以根据需要折叠和穿透。而且，SCIP为整个房子提供了一个R-40

that was more cost-effective to use than cast in place formed concrete walls. Conceptually, the structure was developed as an exo-skeleton against the earth with the inside forms and spaces contained by light gauge cold-rolled steel, which can be folded and penetrated as necessary. The SCIP panel construction provides a R-40 average insulation value for the entire house. In addition the SCIP panels are made from 100% recycled and post consumer foam and have a 50% fly-ash content in the concrete.

The Vail Grant House distinguishes itself from most conventionally developed projects in that it incorporates energy efficient measures that exceed standard practice, optimize building performance, and ensure reduced energy use during all phases of construction and occupancy. The planning and design emerged from close consideration and employment of passive solar design strategies. These strategies include: locating and orienting the building to control solar cooling loads; shaping and orienting the building for exposure to prevailing winds; shaping the building to induce buoyancy for natural ventilation; designing windows to maximize daylighting; shading south facing windows and minimizing west-facing glazing; designing windows to maximize natural ventilation; shaping and planning the interior to enhance daylight and natural air flow distribution.

Furthermore, the building width was intentionally limited to 15' throughout, reducing spans and simplifying construction. Solar panels placed on the slope behind the house produce enough energy to make this a completely self-sustained building.

1 CARPORT
2 ENTRY COURT
3 ENTRANCE
4 MASTER BEDROOM
5 BEDROOM
6 BATH
7 LIVING ROOM
8 KITCHEN
9 DINING
10 WC
11 COVERED PATIO
12 TERRACE

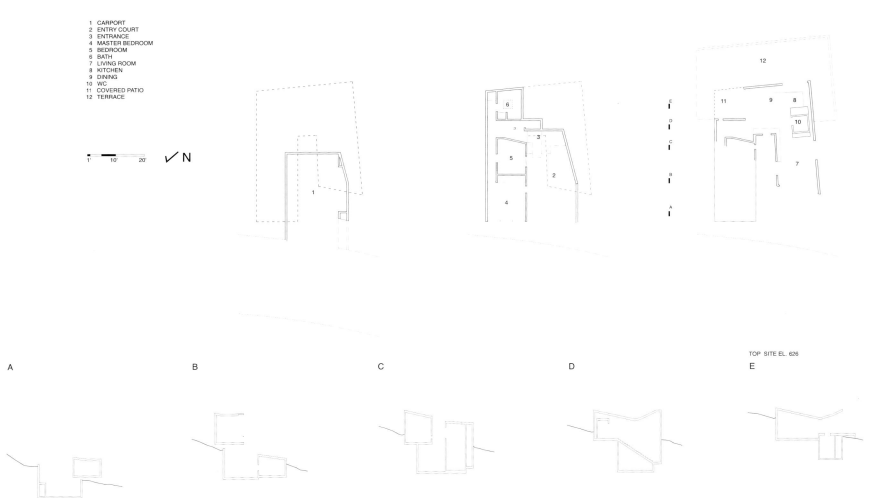

TOP SITE EL. 626
BOTTOM SITE EL. 544

Small diameter pipes are inserted into the hillside and thru the SCIP panels bringing 60 °F air into the building for natural air-conditioning. The large viewing window also allow the sun to heat the adjacent interior concrete slab creating natural convection that rises up the interior space and is vented at the uppermost portions of the building. This will allow natural airflow for both natural cooling and solar radiant heating. When completed the Vail Grant Residence will be 100% energy independent.

By responding to the visceral aspects of the site, both physical and regulatory, a unique sustainable and economic design was achieved.

平均隔热值。SCIP还是由100%可回收和再生泡沫材料制成，并且含有50%的煤灰。

韦尔住宅和传统项目不同之处是：它结合了超出一般标准的节能方法，优化建筑性能、保证在施工和居住阶段减少能源的使用量。这种计划和设计来源于周密的考虑和利用太阳能的策略，其中包括：利用建筑的方向来控制太阳能冷却负荷；利用建筑的定位和定型引入和使用自然通风；设计能够将光照最大化的窗户；设计南面的遮光窗户和减少光照的西向窗户；设计能最大限度保持通风的窗户以及能加强光照和自然空气流通的室内空间设计。

另外，建筑的宽度被特意限制为15英尺，减少了跨度以利于施工。置于房子后面山坡上的太阳能板能够提供足够的能量，以增强住宅能源的可持续性。

小直径管道被插入山坡，通过SCIP板带来60°F的空气，成为住宅的自然空调。宽大的窗户也使阳光温暖内部的水泥板层，形成自然的内部空间上升对流气体，再从最上方的部分排出。这种自然的气流可以引导自然冷热的自我调节。韦尔格兰特住宅是一个100%的能源独立项目。

无论是遵守自然条件还是人为规定，项目都达到了一个独特的可持续性和节省开支的设计目的。这和设计师对建筑所在地内在特点的了解交流不无关系。

Above: site elevation plan
建筑地形剖面图

Contents

Introduction with Article Review, Page 7.

Project Survey

Colorado Court Page 23. COop Editorial Page 41. Nascent Terrain Page 59. Diva Page 67. Orange Grove Page 77. North Point Page 85. Jigsaw Page 95. Solar Umbrella Page 109. SM College Student Service Center Page 117. Bergamot Loft Page 125. Vail Grant House Page 139.

The Firm Page 149.

XAP Page 159. Reactor Films Page 171. Chronology Page 181 & Bibliography.

The Firm

音乐电影制作公司
贝弗利希尔斯，加利福尼亚州

Firm项目是把一个商业建筑空间改建成一个音乐电影制作公司的总部。改造后的位于四楼的办公空间将包括总监和执行小组的办公室以及助理和公司其他员工的办公区，还包括两个会议室，一个厨房和休息室，接待和多功能服务厅。

客户要求从开始设计到项目完成的周期为16个星期。这意味着用于设计构思的时间极其有限。为了达到这个要求，项目实施团队建立了一个系统化的工作策略。在客户、承包商和建筑师空前的共同协作下，建筑终于诞生于创新而有活力的氛围中。

项目根据施工的时间阶段而划分为不同区域来分别执行：每一部分都进行了深度探索，具体的开发策略与客户和承包商共同协商。设计方案的确定也极大地征求了承包商和技术专家的意见。整个执行过程都极其重视时间、造价、设计、施工和制造等一系列因素间的关系。为了顺利地在预算内按时完成项目，所有的图纸都作为项目中的客户演示和施工文件。这种类似于"现场直播"的表达方式使得建筑成为一个直观性的整体画面。

该建筑反映出了客户所具有的一个为音乐和电影工业中的创新者提供舞台的公司的前卫精神。事实上，这个建筑成为一个镌刻在空间和形式内外的精神标志。这个公司致力于在其内部创造一个激励员工之间、上下级之间相互交流沟通的空间环境。创新能量的自由流动和思想交流的良好氛围被认为是公司发展和人才成长的重要因素。另外，当公司的行业准则和工作方式逐渐完善时，他们需要永远保持开放和创新的精神。

The Firm

Location: Beverly Hills, CA
Client/Owner: Jeff Kwatinez

Project team: Lawrence Scarpa, principal-in-charge, Jackson Butler, Byron Merritt, Peter Borrego, Chantal Aquin, Kelly Bair, Nicole Cannon, Heather Duncan, John Jennings, Kirsten Wenck, Gwynne Pugh.

Contractor: Crommic Construction Corp.

Program:

To transform an existing commercial interior space into the signature headquarters for a cutting edge artist management company. Located on the 4th floor of a Beverly Hills high-rise, the program includes executive suites for the principals of the company, offices for other members of the executive team, assistants, and support staff, two conference rooms, a kitchen and lounge area, reception and miscellaneous service rooms.

Solution:

The clients for this project needed to occupy a finished space within sixteen weeks from the beginning of the design process. This allowed for a very limited incubation period in which ideas had to be generated and realized. To meet this demand, a systematic working strategy that capitalized on the extreme constraints of the project and a team approach was adopted. Client, contractor and architect collaborated with an unprecedented synergism. Ultimately, the architecture emerged from the inventive and dynamic atmosphere that was cultivated in response to the particular conditions of the project.

The program was strategically divided into distinct and separate areas that could be developed and detailed in phase with the construction schedule: each programmatic element or area was explored in depth and developed in detail, presented to the client and then dimensioned and issued to the contractor for construction. Design decisions were made in close association with the contractor and various fabricators whose expertise was fundamental to the project. A complex set of issues and relationships involving time, money, design, construction and fabrication created a context in which emphasis was necessarily placed on the process of making and the craft of construction. In order to realize the Firm on time an on budget, all drawings generated for the project served as both client presentation and construction document. The immediacy of working in this 'one take' or 'live broadcast' context resulted in an architecture that, in essence, evolved as a drawing at full scale.

Below: interior view of the Firm

室内

The architecture that emerged from this project reflects the spirit of the client for whom it was created. In essence, the architecture becomes a signature—etched into an out of space and form—for this cutting edge company that promotes and supports the careers of fresh and innovative talent in the music and film industries. The Firm is committed to creating an environment in which cross-pollination within the company can occur. Hence, they wanted a space that would encourage interaction amongst employees and across hierarchies. A free flow of creative energy and sharing of ideas is deemed critical to the well being of the company as well as to the growth and development of the emerging talent whom they represent. Furthermore, while the company operates within a field that already has well established industry standards or ways of working, they maintain and open and inventive spirit at all times.

The architecture responds by creating a landscape that bridges the film and music 'communities' of the company. Core elements of the existing structure are stripped bare and maintained as established fixed order. New insertions of space and form flow in and around the existing container shaping a dialogue between the old structures optimized to its purest condition and these new inventive forms. The juxtaposition of these formal hierarchies creates a dynamic tension that enhances the spatial experience of

Right below and opposite:
interior view of the Firm

室内

该建筑的设计也对应了公司在电影和音乐两方面的专业结合。建筑原有的核心结构元素被剥离裸露出来，形成一种凝固的次序。新插入的空间和形式荡漾在旧的空间中，架构了新形式与质朴的原有结构之间的对话。这些形式间的碰撞创造了一个动感的张力，增强了两者的空间体验。虽然公司的电影和音乐部门占据着空间的不同区域，但是两者之间并没有特别的界限。它们在空间上、在视觉上被公共接待区和休闲室联系在一起。被多种开放元素围绕的中心会议区更起到一个广场的作用，在这里客户能够闲逛、看电视、社交。在这个公共中心地区，交流和会议可以在更私人化的办公室中自由进行。这个设计旨在打破公司中通过建筑布局形式所体现出的等级观念并反对界限和限制。被抛光过的水泥地面形成了一个连续的整体平面。这个平面联系着不同的流动元素；这些元素激活着空间，并有节奏地引领整个空间，与其中裸露的灰色结构柱达成了平衡。

形式元素冲破了界限的禁锢。其表面脱离了裸露的空间层次和肌理，层叠的墙壁在空中凸现，桌子台面变成了座垫，复杂多样的形式刺激了观众也激活了空间。这种交迭延伸的语言引发了一种充满活力的潮流，并建造了一个视觉长廊，它能让视线穿越空间并最终加强其联系和连续性。就是在这样的运动中，新老事物之间的交流才如此地明显。新元素的加入引起了对原有空间内在素质的认知，而原有空间也为人们对新形式的开发和欣赏提供环境。

both. While the film and music divisions of the company occupy distinct areas in the space, there is never a decisive division between the two. They are connected spatially and visually by the more public reception and lounge area of the Firm. This central meeting zone, around which all other elements unfold, begins to function more like a piazza or public square where clients can hang out, watch TV, mingle etc. Conversations and meetings can freely flow from more private offices into this more public central zone as necessary. Architectural elements reinforce this concept of flowing space. While hierarchies within the company are architecturally expressed (i.e. the executives occupy distinct and central offices on the edge of the building) the architecture works to break down and confront boundaries and containment. The polished concrete slab at the ground plane establishes a continuous field that remains constant throughout. This plane serves to unite the free floating; programmatic elements that animate the space and furthermore bring them into a balanced composition with the exposed grid of structural columns that rhythmically order the overall spatial container.

Formal elements defy planar boundaries. Surfaces peel away exposing layering of space and textures. Wall folds to become overhead projection. Tabletop becomes seating; these formal complexities confront the viewer and animate the space. A language of overlap and extending planes stimulates a dynamic flow and constructs visual corridors that carry the eye across the breadth of the space ultimately reinforcing connectivity and continuity. It is in these moves that the interaction between the old and the new becomes evident. New insertions promote an awareness of the inherent qualities of the existing container and the existing container provides the landscape in which new forms can be explored and appreciated.

Contents

Project Survey

Introduction with Bicycle Review, Page 7. Colorado Court Page 23. COop Editorial Page 1. Nascent Terrain Page 59. Diva Page 67. Orange Grove Page 77. North Point Page 85. Jigsaw Page 95. Solar Umbrella Page 109. SM College Student Service Center Page 117. Bergamot Loft Page 125. Vail Grant House Page 139. The Firm Page 149. **XAP** Page 159. Reactor Films Page 171. Chronology Page 181 & Bibliography

XAP

Below: interior view of XAP, the conference room
XAP 室内和会议室

XAP

Location of Project: Culver City, California
Client/Owner: XAP Corporation
Total Square Footage: 22,000 sq. ft.

Project team: Lawrence Scarpa - Principal- in-Charge. Kelly Bair, Peter Borrego, Angela Brooks, Michael Hannah, Vanessa Hardy, Anne Marie Burke, Anne Marie Kaufman Brunner, Ching Luk, Tim Petersen, Gwynne Pugh, Bill Sarnecky, Lawrence Scarpa.

General Contractor: Hinerfeld Ward Inc.
Photography: Benny Chan, Fotoworks

The design of this 22,000 square foot tenant improvement for XAP Corporation evolved from the unique challenges and conditions presented by the client and the exigencies created by the physical context. XAP Corporation is the pioneer in electronic and Internet-based information management systems for college-bound students. XAP Corporation's mission is to be the leader in building and providing students and their families, universities and sponsors the most comprehensive and widely used online information services for higher education.

In addition to particular programmatic needs, the XAP project presented the delicate challenge of operating within and conforming to distinct parameters issued by the building's architect and owner/developer. All phases of design required approval via rigorous design review by not only the client but, also more uniquely, the building's architect. One of the more stringent constraints required by the building architect was that no element of the new tenant improvement design significantly touch or interact with the existing structure. This imposed requirement significantly influenced the resulting formal strategy and design approach.

The organizational strategy implemented at XAP is one frequently revisited by Pugh + Scarpa. Offices and workstations, which constitute a dense area of the program and require similar spatial properties (size and shape) and formal elements (desk, chair, shelves, etc.), are clustered and neatly organized in an open landscape. Service spaces and additional offices requiring more privacy are organized in simple volumes that flank the perimeter of space. These more private areas and individually inhabited areas of the program deliberately serve as a background for a more dramatic expression of the public space and selected areas of the program which can then be foregrounded, emphasized and more dynamically explored. Free from full height, enclosing walls, the open landscape strategy supports a continuity and flow of space. Even though the 22,000 SF building is packed with program, it maintains an open, spacious feeling. This organizational strategy also takes advantage of the spatial qualities of the existing building—an industrial saw-tooth roof warehouse with exposed framework and dramatic clerestory windows at each structural bay. Finally, the placement of formal bars of program at the perimeter allows for the creation of an elegant circulation space through which the buildings mechanical and electrical services are quietly distributed. Each bar is characterized by a tall, clean, simply expressed soffit behind which the building's infrastructure pulses. This design solution was particularly significant as the building's architect required concealment of all ductwork and electrical conduit.

While designed to serve as formal background, these workstation and office elements are well considered, meticulously crafted and elegantly realized. Furthermore, they establish a rigorous order that creates an ever-present backbone that can then be broken from and contrasted by the freer flowing, more sculptural elements without losing the clarity and coherence of the overall design.

With this well-constructed backbone in place, the entry and public areas of XAP unfold along the southern edge of the building. Flooded with light from clerestory windows, this is the zone where forms play and space dances. Unlike most offices, the reception desk at XAP is held back from the main entry door by 30 feet. This encourages

XAP项目
卡尔弗城，加利福尼亚州

XAP是一个占地22 000平方英尺的建筑，它在适应客户要求和对物理环境的急迫挑战中得以形成。XAP是为高校学生提供电子和互连网信息管理系统服务的先锋企业。它的目标是成为建筑业的领袖，为学生以及他们的家庭、大学和赞助者提供最综合和最广泛的高等教育在线信息服务。

除了特殊的规划需要，XAP项目面对了操作过程中因建筑师和业主及开发商各方因素所引发的问题的挑战。所有的设计方案都要求由客户和建筑师共同审核批准。建筑师面临的一个更严格的限制是新的设计元素绝对不能触及或者破坏现有建筑的结构。这个强制性的要求极大地影响了最终的策略和设计方案。

XAP的实施策略不断被皮尤+斯卡帕修订。一个密集的办公和工作站区域组成所需要的类似的空间道具（大小和形状）以及形式元素（书桌、椅子、书架等）都整洁有序地排放成一个开放景观。在公共空间的侧翼安排了服务空间和非常私人化的办公室。这些非常私人化的区域映衬出了公共区域的戏剧化表现，它作为前景突显了空间的活力。虽然这个22 000平方英尺的建筑被各种功能区域充满，但它保留了一个开放广阔的感觉。这个组织性策略充分利用了现有建筑的空间特性———一个具有锯齿状屋顶的、无掩蔽的框架和活动天窗的工业仓库。最后，机械和电子设备控制系统有序地置放在建筑周围，并创造了一个优雅的空间流通环境。每一个设备系统都安置在建筑原有的高大、明确简约的拱腹结构背后。这个设计出色地解决了建筑师提出的在建筑中隐藏所有的通风和电子设备管

Above: conference room view

会议室

道系统的要求。

这些工作站和办公设备作为空间形式背景，都被仔细考虑、精巧制作和优雅地实现了。另外，它们也确立了一个严格的次序，形成一个永久性的空间骨架，在拆解和自由流动的状态下，将维系整体设计的一致性和意义的明晰。

在这个基础上，XAP的入口和公共区由建筑的南边展开。明亮的光线从这里的天窗洒进，这是形式展现和空间起舞的地方。与大多数办公室不同，XAP的接待桌离正门有30英尺，这使得来访者到这里首先参与的是空间交流。他们在身体感观上直接体验公司的风格和特点，而不是被隐藏在幕后的公司形象。XAP的公司形象由此被充分展示在人们面前，毫无保留地给人真正的空间体验。

皮尤+斯卡帕设计了XAP的大多数家具，这是整个项目不可缺少的一部分。作为大型的设计元素，工作站、会议桌、沙发、椅子和地毯将对空间的效果起重要影响。 XAP特色家具之一的接待桌在形式上表达得很充分，制作也十分精良。它悬于钢管支架之上，有两个点和地面接触。虽然看上去很复杂，但事实上制作起来非常简单：混凝土和焊接钢管的结合。XAP的家具是精美工艺品和空间的双重体验，这些元素的形式和表现是激活和创造空间良质的不可替代因素。

建筑中最繁忙的区域是厨房及咖啡厅和娱乐区。这些地点被安置在这个30英尺×145英尺的回廊空间对面的尽头，一个充满活力的连接线穿越了整个建筑。这打破了传统企业的办公室旧模式——把厨房设计在后部无法看见的房间内，也没有娱乐区。人们必须走过这个回廊才能进入上述区域去体验这个

visitors to penetrate the space and interact with form before making contact. They are invited to experience the attitude and character of the firm physically and sensually rather than abruptly being hindered at an anonymous waiting area behind whose walls a corporate identity is concealed. XAP's corporate identity unfolds up front and unabashedly as a spatial experience.

Pugh + Scarpa designed much of the furniture at XAP as an integral part of the project. Workstations, conference tables, sofas, chairs and carpets are as critical to the spatial effect as the larger elements of the design. The reception desk is one of the feature furniture elements at XAP. Formally expressive and meticulously crafted, the reception counter floats atop its tube steel support, touching the ground at two points. While seemingly complex, the reception desk is actually composed of quite simple construction methods: cast-in-place concrete and welded tube steel. While it can be experienced as a finely crafted object, the reception desk, as well as many of the other design elements and furniture at XAP, can also be experienced as space. The form and expression of these elements activate and create a quality of space that would not exist without them.

The busiest, most highly trafficked areas of the program—the Kitchen/Cafe and the Recreation Area—are placed at opposite ends of this 30' x 145' corridor of space. This organization creates a dynamic link across the length of the building and defies typical corporate office patterns of sticking kitchens in a room in the back, not to be seen, while not providing recreation areas at all. One must walk clear across the space to get to either of these programmatic elements and hence, experience the space and its sculptural follies in the round. The kitchen is dominated by a 20' long cast in place concrete island that serves as a bar counter and as another node of activity.

Carefully oriented, the Board Room and the Conference Room become the formal focal points of the project. Both of these organically shaped forms are representative of an ongoing formal exploration and signify an important aspect of the firm's work: a commitment to experimentation with form and materials and a passion for the process of making and construction. These qualities are as integral and important to the design as the resulting form itself. These sculptural follies were initially studied in physical model form. From there, a series of computer generated models were produced to study the formal possibilities, as well as location, orientation and relationship to the larger field. Once the final design was achieved and engineered, the computer models were used to generate drawings for a steel bending fabricator who used the data to fabricate the full scale

空间和它周围的景观。厨房中心设置了一个20英尺长的水泥砌筑的岛状烹台，它也是吧台和举行其他活动的中心。

董事会会所和会议室是项目的设计焦点。这两个部分都是持续进行的具体形式探索，并意味着设计师的工作状态：对形式和材料的实验承诺，以及对制造和建设过程的热情。这些素质对设计和形式结果本身来说是整体的和重要的。这些雕塑物起先是以物理模型的形式体现的，之后，一系列电脑模型被制作出来以研究形式的可能性、位置、方向和与整个空间的关系。当最终的设计被确定和决定实施后，电脑模型生成的草图和数据交由钢筋制作商加工制作，每一根被组装起来的预制钢管都被赋予了生命。这些钢管结构都被放置并暴露在物体的外面。外露的钢管和周围的系统组件使形式的表达极显特色。内部由石膏板拼接完成，非常光滑而且精致。这种材料和肌理主导了XAP的整体设计风格，增强了空间的活力和华美度。

在大多数办公室空间里，公用电话通常设在靠近休息室角落的墙上。为了更进一步激活XAP建筑的循环和体验，公用电话被方便而聪明地设计在董事会会所和会议室尽头的折角拐弯处，毫不刻意地融合到整个建筑形式中，那些寻找私人对话空间的员工对这些电话角非常满意。

XAP项目还有另一个空间组织上的挑战。虽然公司希望有一天能够占领整个空间，但他们需要留出6000平方英尺的面积作为独立的转借或出租空间。这部分空间的设计要考虑它的临时性和未来有可能作为XAP公司总部一部分的特点。

无论是物理、财务、法规和规划还是所有上述所有

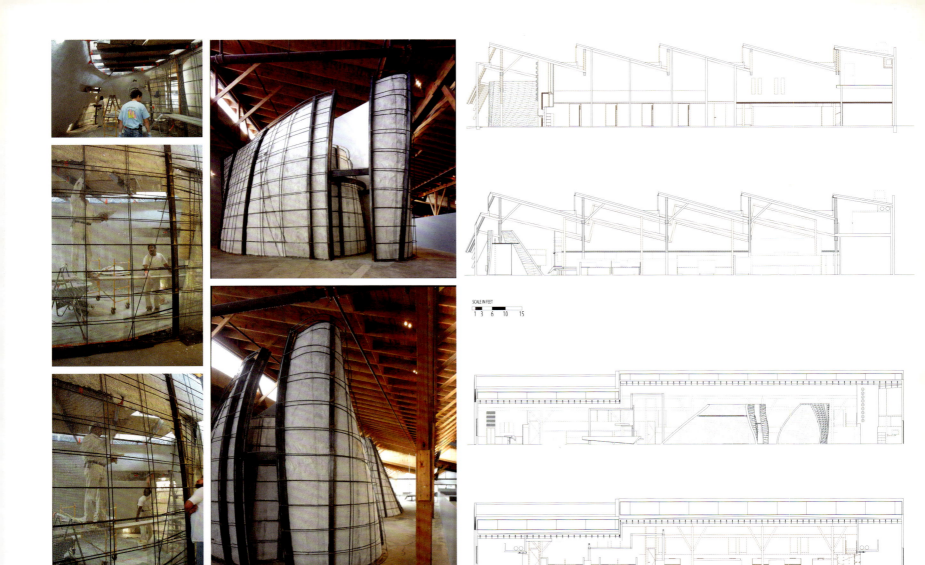

因素，在一个项目中它们通常都是不确定的。参与XAP项目的一个授权机构在最后一天增加了挑战性的限制，但这更加激励了建筑师在满足客户对形式和功能要求上的持续探索，这同时也是设计事务所工作和文化发展的助燃剂。

Above: construction view and interior view
施工现场和室内

Right above: house section
建筑剖面图

Below: house detail
室内细部

tube steel members which, ultimately assembled, give presence and life to each volume. Significantly, these structures were left literally exposed and raw on the exterior. Tube steel and round bar characterize the formal expression. In a twist of convention, the interior is finished with plaster—smooth and refined. This material and textural counterpoint is indicative of the overall design at XAP, enhancing the dynamism and richness of the space.

In most office spaces, public telephone booths are conventionally located on a wall near the restrooms or in the back corners of a space. To further activate circulation and the experience of the architecture at XAP, the booths are conveniently and cleverly contained in the warps and folds at the tail end of the board room and conference room. Seamlessly attached to the forms, these telephone rooms are greatly appreciated by the staff seeking private conversation space in an environment, which is largely designed around an open plan.

The XAP project presented an additional organizational challenge. While the company hopes to one day occupy the entirety of the space, they needed to maintain a 6,000 SF portion as a distinctly separate subtenant lease space. This space was designed for both its immediate condition as separate tenant space and for its future condition as part and parcel of the XAP corporate headquarters.

There is nothing unusual about constraints in a project, whether they are physical, financial, code, programmatic or all of the above. These are the usual suspects. The XAP project's introduction of an additional, highly involved authorizing agency merely enhanced the challenge and at the end of the day, stimulated the ongoing quest to arrive at a process and design that satisfies the client's formal and functional needs while also fueling the growth and culture of the firm's work.

Opposite: interior view of XAP

XAP 室内

Contents

Project Survey

Introduction with Article Review, Page 7.
Colorado Court Page 23. COop Editorial Page 41. Nascent Terrain Page 59. Diva Page 67. Orange Grove Page 77. North Point Page 85. Jigsaw Page 95. Solar Umbrella Page 109. SM College Student Service Center Page 117. Bergamot Loft Page 125. Vail Grant House Page 139. The Firm Page 149. XAP Page 159. **Reactor Films** Page 171.
Chronology Page 181 & Bibliography

Reactor Films

Reactor电影工作室项目

圣莫尼卡，加利福尼亚州

该项目是将一座20世纪30年代建造的石材艺术画廊改建成用于电视广告和音乐视频制作的办公和工作空间。

Reactor 项目面临的挑战是：客户要求在14个星期之内完成从设计到施工的任务。为了达到这个目标，我们设计了一个由客户、承包商和建筑师共同协作的充满创意和活力的系统化工作策略。

Reactor 项目被策略性地拆分为几个独立的区域，这些区域能够按照施工阶段来开发和细化：每个规划元素或者区域在经过深度探究和细致开发后展示给客户，然后由承包商进行施工。设计方案是与承包商和制造商密切合作下共同制定的，后者的专业技巧是项目成功的关键。与制作过程和施工工艺相关的时间、造价、设计、施工和制造等一系列复杂问题成为设计的中心要素。项目施工开始于设计的第一个星期并于第二个星期得到市政的许可。所有的图纸都作为项目中的客户演示和施工文件。为了使得这个过程更加容易，并加强参与者之间的沟通，所有的设计图都通过手绘的形式体现在11′x 17″的牛皮纸上，这种类似于"现场直播"的表达方式使得建筑成为一个直观性的整体画面。

在空间上，建筑围绕着一个中心会议室，并设计了一个公共通道。会议室坐落于"通道"大厅，这里置放了一个从长滩码头购回的旧海洋集装箱。对这个陈旧的物件进行有创意的再利用是当时的经济环境下的产物：因为与日本贸易的不平衡，这个旧集装箱的成本非常低廉。罗伯特·文丘里曾经说过："一个原本为我们所熟悉的事物，经由新的

Reactor Films

Location of Project: 1330 4th Street, Santa Monica, California
Client/Owner: Stoney Road Productions and Reactor Films
Total Square Footage: 7000 sq. ft.

Project team: Lawrence Scarpa - Principal- in-Charge. Angela Brooks, Jackson Butler, Adam Davis, Mike Ferguson, John Jennings, John Mulcahy, Tim Peterson, Gwynne Pugh, Sharon Robertson-Bonds and Lawrence Scarpa.

Furniture and Fixture Design: Mike Ferguson, John Jennings and Lawrence Scarpa (with Dave Scott) .

Structural Engineering: Gwynne Pugh and Sharon Robertson Bonds.

Steel and Furniture Fabrication: Dave Scott of DESU.

Construction Team: Brian Crommie and Tom Hinerfeld of BT Builders.

Consultants: Gwynne Pugh - Structural Engineering; Dave Scott - Steel and Furniture Fabrication; Richard Godfrey - Light Trough

General Contractors: BT Builders, Inc.
Photography: Marvin Rand

Program: To remodel an existing 1930's Art deco Masonry Building Art Gallery into office and work space for production of TV commercials and music videos.

Solution: Reactor presented the unique challenge of satisfying the client's requirement to move into a completed space in less than fourteen weeks from the beginning of the design process. In order to meet this demand, a systematic working strategy was developed to capitalize on these extreme constraints while cultivating an inventive and dynamic working atmosphere in which client, contractor and architect collaborated with an unprecedented synergism.

The program was strategically divided into distinct and separate areas that could be developed and detailed in phase with the construction schedule: each programmatic element or area was explored in depth and developed in detail, presented to the client and then dimensioned and issued to the contractor for construction. Design decisions were made in close association with the contractor and various fabricators whose expertise was fundamental to the project. A complex set of issues and relationships involving time, money, design, construction and fabrication created a context in which the process of making and the craft of construction intensified in importance and became central aspects of the design process. Construction commenced during the first week of design and permits

Above: shipping container
海运集装箱
Right: interior view of Reactor Films
Reactor 室内

Opposite: interior view of Reactor Films, the office box

Reactor 室内，办公室中心箱体

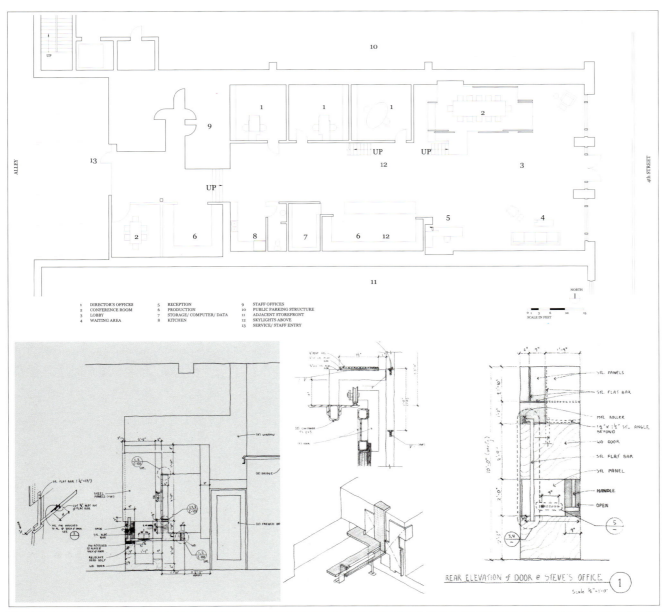

1 DIRECTOR'S OFFICES	5 RECEPTION	9 STAFF OFFICES	
2 CONFERENCE ROOM	6 PRODUCTION	10 PUBLIC PARKING STRUCTURE	
3 LOBBY	7 STORAGE/ COMPUTER/ DATA	11 ADJACENT STOREFRONT	
4 WAITING AREA	8 KITCHEN	12 SKYLIGHTS ABOVE	
		13 SERVICE/ STAFF ENTRY	

Reactor Films

Opposite: interior views
室内

Left upper: house plan and details
平面图和细部
Left bottom: interior view
室内

Reactor Films

Below: house detail and exterior view from street

建筑室内细部和室外

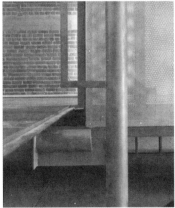

were issued by the City by the beginning of the second week. All drawings generated for the project served as both client presentation and construction document. To facilitate this process and allow for rapid facsimile communication between participants, all drawings were completed freehand on 11' x 17" vellum. The immediacy of working in this 'one take' or 'live broadcast' context resulted in an architecture that, in essence, evolved as a drawing at full scale.

Spatially, the project revolves around a centrally located conference room, positioned to engage the public street. The conference room, located in the "street" lobby, re-occupies a used ocean shipping container purchased from the Long Beach shipping yard. The economic climate at the time of this project permitted the inventive reuse of this ready-made object: Because of the trade imbalance with Japan, the used container was readily available at an extremely low cost. Robert Venturi once said, "A familiar thing seen in an unfamiliar context can become perceptually new as well as old." Like the 1930's building that this project occupies, the recycled container is transformed and perceptually repositioned to capitalize on its inherent history. In essence, it exhibits a spatial biography, its surfaces and voids charged with fragments of memory etched into it over time. The surrounding interior space was

Below: stair, interior view, container sitting

楼梯，室内和海运集装箱安置图

conceived as a fluid surface wrapper rotating asymmetrically around the centroid of the container. This surface wrapper alternately pushes close to and peels away from the walls and structure of the existing building. This push and pull or concealing and revealing formal strategy suggests a dynamic relationship between the new and old while indicating a design attitude that respects the integrity of the old while maintaining a commitment to the generation of an inventive and thoughtful new. While the surface plane remains flat and orthogonal, it is consciously exploited for sculptural expressiveness. Molded into an extra dimension in its wrapping, the voids which it creates become as important as the surfaces themselves. Walls, rooms, and windows create singular experiences yet balance in tension together as a cohesive composition. Ultimately, Reactor is an attempt to stimulate meaningful experience in architecture through the process of making. It becomes a questions of "how" rather than "what".

方法改造后，会带给人新老并存的感觉。"就像这个20世纪30年代的老建筑，这个被再利用的集装箱改造了它，并且在感觉上重新定位了它内含的历史。在本质上，它展现了一个空间纪年，它的内外都充满了时间记忆的碎片。建筑的内部空间环境被构思成一个流动的外包装，它围绕着集装箱中心不对称地循环。这个外包装交替地靠近并离开现存建筑的墙壁和结构。这种一推一拉或者说是隐藏和显现的形式暗示了新旧事物之间的动感关系，也表明了一种在尊重原有事物的重要性的同时，也注重创意和新思维的设计态度。表面体现出来的平面直交的形式是一种刻意的雕塑式表现。外包装被塑造于外部范围，由此而产生了和表面本身同样重要的一个空间。墙壁、房间和窗户创造了风格各异的感受，同时也结合在一起，通过张力达到共同平衡。最终，Reactor 项目成为一种通过制作过程来激发深刻感受的尝试。它成为一个有关"如何"而非"什么"的问题。

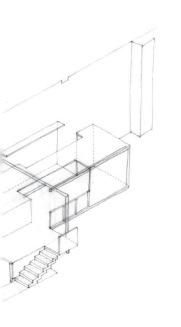

Contents

Introduction with Article Review, Page 7.
Project Survey Colorado Court Page 23. COop Editorial Page 41. Nascent Terrain Page 59. Diva Page 67. Orange Grove Page 77. North Point Page 85. Jigsaw Page 95. Solar Umbrella Page 109. SM College Student Service Center Page 117. Bergamot Loft Page 125. Vail Grant House Page 139. The Firm Page 149. XAP Page 159. Reactor Films Page 171.
Chronology Page 181 & Bibliography

Chronology

Lawrence Scarpa, AIA
Principal

Mr. Scarpa is the principal-in-charge of design at Pugh + Scarpa and has been practicing architecture since 1989. While pursuing his personal practice, he is also an educator in design and construction technology with a special emphasis on sustainability.

While working in the offices of Gene Leedy, Paul Rudolph and Holt, Hinshaw, Pfau & Jones, he was involved in the design and construction of over twenty projects, which have received international, national and/or local awards. In 1997, the Academy of Architecture Arts and Sciences named Mr. Scarpa as one of the top 39 architects worldwide under age thirty-nine. Mr. Scarpa's commitment to building sustainable communities is demonstrated by his success in garnering public and private support for his projects. Mr. Scarpa received a $75,000 grant from the United States Department of Energy for implementation of the Electric Vehicle Charging Station, a demonstration project located at City Hall in Santa Monica. He also received a total of $500,000 in funding (with Dr. John Ingersoll) from the United States Department of Energy, the California Energy Coalition, the Southern California Gas Company, the Regional Energy Efficiency Initiative and the City of Santa Monica to develop and implement energy efficient strategies for Colorado Court.

Complementing his professional activities, Mr. Scarpa has taught and lectured at a broad range of institutions nationally and internationally. He is a co-founder and current board member of Livable Places, Inc., a non-profit development company. Livable Places' mission is to provide more livable and sustainable affordable housing on problematic urban sites and to influence and change the vision of urban policy makers and voters. Livable Places recently received nearly $1,000,000.00 in grants from The Irvine Foundation, Fannie-Mae Foundation, Washington Mutual and California Federal Bank. He is a co-founder of the A+D Architecture and Design Museum Los Angeles, serves on the editorial board for LA Architect magazine and the Board of Directors for the AIA/LA. In 2002, Mr. Scarpa was the jury chair for the National AIA Interior Design Awards Program and the State of Iowa and State of Oregon AIA Design Awards Programs. He recently received a $50,000.00 National Endowment of the Arts grant to sponsor

national design competition for Livable Places. Over the last five years Mr. Scarpa has won twenty major design awards including seven National AIA Honor Awards, the 2003 Rudy Brunner Award and the AIA COTE "Top Ten Green Building" Award. He is a finalist for the 2003 World Habitat Award and his work will be exhibited at the National Building Museum in Washington, DC in 2004. He was selected in 2004 by the Architectural League in New York as an Emerging Voice in Architecture.

Education
1981
Bachelor of Design,
University of Florida

1987
Master of Architecture,
University of Florida

Registration
1989
NCARB Certified
Registered Architect, California C21812

2000
Registered Architect, North Carolina 8534

1988
Registered Architect, Florida, AR 00132227

Affiliations
American Institute of Architects Los Angeles Chapter, Board of Directors

Society of Architectural Historians, Life Member

The National Association of Educators, Former Member

Southern California Association of Non Profit Housing

Livable Places, Inc., Board of Directors (Co-Founder)

A+D Museum, Los Angeles, Co-Founder

2002-2004
LA/AIA, Board of Directors

2000 – present

LA Architect Magazine, Editorial Board,

CityWorks, Advisory Board of Directors

Lectures

2004
Keynote Speaker, AIA Colorado Conference
Master of Architecture series, Los Angeles County
Museum of Art, Catholic University
AIA Georgia
Architecture League of New York, "Emerging Voice"
Woodbury University
University of California Berkeley
AIA National Convention, Chicago (with Peter Davey,
Editor, Architectural Review)
California College of the Arts, San Francisco
Keynote Speaker, AIA Baltimore Chapter Conference
and Green Week
Keynote Speaker, State of Florida AIA State
Conference
National Building Museum, Washington, DC
Northwest and Pacific Regional AIA Conference,
Oregon
Keynote Speaker, Southwest Regional AIA
Conference, Baton Rouge, LA

2003
Harvard University
AIA Young Architects Forum, Kansas City
ACSA Sustainable Pedagogies Conference,
Cranbrook Academy
Keynote Speaker, Critical MASS, University of North
Carolina, Charlotte
University of Southern California
Mississippi State University
Arizona State University
University of Texas at Austin
University of Texas at Arlington
State of Arizona Historic Preservation Convention
AIA, Tucson Chapter
AIA, Portland Chapter
Keynote Speaker, Monterey Design Conference
University of Southern California
AIA National Convention, San Diego
AIA Young Architects Forum, Los Angeles
Green Strategies Conference sponsored by
Mississippi Valley Gas
Alternate Energy Conference sponsored by Los
Angeles Dept. of Water and Power

2002
University of Colorado, Boulder
Seminar Speaker, International Green Building
Conference, Austin, Texas
Keynote Speaker, State of Iowa AIA Convention
Lecturer "Warm, Dry and Noble: The Philosophy of
Samuel Mockbee" UCLA Hammer Museum
Seminar Speaker, 2002 National AIA Convention,
Charlotte, NC
Pratt Institute, Brooklyn, NY
Tulane University, New Orleans
SCI-arc
AIA National Convention, Charlotte

2001
California Polytechnic University, Pomona
Woodbury University

2000
California Polytechnic University, Pomona
Woodbury University
University of Florida

1999
University of New Mexico

1997
SCI-arc

1991
Mississippi State University

Awards Juries
2004
Young Architects Forum "Monsters of Design"
Kansas City

2004
Southwest Regional AIA Awards, Jury Member

2004
State of Oregon AIA Awards, Jury Member

2004
Northwest and Pacific Region AIA Awards, Jury
Member

2003
National AIA Interiors Awards, Jury Chair

2003

State of Oregon AIA Awards, Jury Member

2002

State of Iowa AIA Awards, Jury Chair

2001

State of Arizona Masonry Institute, Jury Member

Academic Positions
Friedman Professor in Architecture, University of California, Berkeley
SCI-Arc
Otis College
University of Florida
Mississippi State University
Woodbury University

Gwynne Pugh, AIA, P.E.
Principal

Mr. Pugh has been practicing architecture, planning, civil and structural engineering since 1971. He is responsible for general project administration, technical and production management of the projects.

Mr. Pugh's architectural, planning and engineering experience covers an extensive range of projects. These projects have included replacement and new affordable housing, retrofit and adaptive re-use of structures for varying uses. He has extensive experience working with community groups and public and private agencies and is currently in charge of the Courson Connection Master Planning project for the City of Palmdale. He serves on the Santa Monica Planning Commission, as a Peer Review consultant to the City of Carson and with the City of Los Angeles and the Getty Conservation Institute.

Mr. Pugh started practice in a consulting engineering firm of Posford Pavry & Partners in Central London where he worked on projects for Ford Motor Company and on an historic greenhouse at Kew Botanical Gardens. After completing his engineering degree at Leeds he received a three year scholarship to study architecture at UCLA. Concurrently he worked at Wynne Engineering where he was involved in land development and civil and structural engineering design throughout California. While working at Bardwell, Case, & Gilbert he worked on residential and mid scale commercial projects He was project manager for a 100,000 square foot conference and entertainment center at the Disneyland Hotel for I.K. Weber. He has subsequently designed projects for the Disneyland Hotel, Seven-Up, Coca-Cola, the Getty Conservation Institute, Santa Monica College, the City of Santa Monica, City of Carson, City of Upland, City of Palmdale, City of Los Angeles, City of Alhambra, County of Los Angeles, Los Angeles Community College District as well as numerous other cities and Institutions.

Mr. Pugh is a LEED™ Accredited Professional, and is considered an expert in sustainable structural design and engineering.

Education
1978
Master of Architecture,
University of California Los Angeles
School Architecture and Urban Planning

1975
Bachelor of Science,
Architectural Engineering
Leeds University, Great Britain

Registration
Professional Engineer, California C33452
Registered Architect, California C14937

Affiliations
American Institute of Architects
American Society of Civil Engineers
Structural Engineers Association of Southern California

Angela Brooks, AIA, LEED™ Accredited Professional
Principal

Angela Brooks, AIA has been practicing architecture since 1991. She has extensive experience working with local public agencies on a variety of institutional projects. Most recently, Ms. Brooks serves as the consultant for LEED Certification and Commissioning on the TreePeople Center for Community Forestry. She has served as Project Manager on several projects at Santa Monica College and projects within the cities of Santa Monica and Los Angeles, including

Colorado Court, Step Up On Fifth, 1424 Broadway and Fuller Lofts.

Ms. Brooks is considered a leader in the field of environmental and sustainable design and construction. She has lectured extensively on the topic of sustainability, including the 2003 AIA National Convention, The 2002 USGBC (United States Green Building Council) National Convention and a host of other national events. She is currently working with many organizations and design professionals as a sustainability consultant.

Ms. Brooks worked in the offices of Skidmore, Owings & Merrill and Hellmuth, Obata & Kassabaum, where she served as a member of the design team on large commercial projects that included the Tampa Convention Center, Kaiser Permanente hospital, and Symphony Towers, a high-rise office building in San Diego. After completing her formal education, Ms. Brooks worked in the office of Appleton, Mechur & Associates on large-scale multi-family housing projects. Her work has won numerous accolades including first prize in The New Public Realm International Design Competition, PA Award, National AIA Design Award, and publication in Dense-City Lotus International documents in 1998. From 1993 to 1996, Angela worked at the Los Angeles Community Design Center, a private non-profit development company dedicated to building affordable housing, community and senior centers and to create stable neighborhoods. Her responsibilities ranged from project architect to construction manager on a variety of housing projects, which included adaptive re-use of historic structures, remodeling of existing overcrowded apartment buildings and construction of new community based projects.

From 1996 to 1999, Ms. Brooks worked in the office of Killefer Flammang Purtill Architects as an Associate in the firm. Her work included a private high school with a 300-seat theater in Las Vegas, and a $4,000,000.00 Santa Monica municipal pool project. She was the senior project architect on the Old Bank District Project, a 224-unit artist loft project in downtown Los Angeles for Gilmore Associates, a private developer. In 1999 Ms. Brooks joined Pugh + Scarpa as a Principal. and is responsible for overall project and staff management. She is a founding member and President of Livable Places, Inc., a non-profit development company dedicated to building sustainable mixed-use housing in the city of Los Angeles on under-utilized parcels of land as a reaction against Southern California's suburban sprawl.

Education
1991
Master of Architecture,
Southern California Institute of Architecture

1987
Bachelor of Architecture,
University of Florida

Registration
1997
Registered Architect, California C27554

Affiliations
American Institute of Architects
Livable Places, President

Pugh + Scarpa

Pugh + Scarpa, has redefined the role of the architect to produce some of the most remarkable and exploratory work today. They do this, not by escaping the restrictions of practice, but by looking, questioning and reworking the very process of design and building. Each project appears as an opportunity to rethink the way things normally get done – with material, form, construction, even financing – and to subsequently redefine it to cull out to latent potentials – as Lawrence aptly describes: making the "ordinary extraordinary." This produces entirely inventive work; work that is quite difficult to categorize. It is environmentally sustainable, but not 'sustainable design;' it employs new materials, digital practices and technologies, but is not 'tech or digital;' it is socially and community conscious, but not politically correct. Rather, it is deeply rooted in conditions of the everyday, and works with our perception and preconceptions to allow us to see things in new ways.

Pugh + Scarpa is an architecture, engineering, interior design and planning firm founded in Santa Monica in 1991. Pugh + Scarpa has grown to a firm of 43 professionals and is currently working on an assortment of commissions for public, private and institutional clients. Pugh + Scarpa maintains offices in Santa Monica, California, San Francisco, California and Charlotte, North Carolina.

Services

Architecture
Engineering
Planning
Interior Design
Sustainable Building, Energy and Cost Benefit
Analysis and LEED™ Certification
Energy Modeling
Building Commissioning
Facilities Planning and Programming
Life Cycle Cost Analysis
Existing Building Evaluation
Financial Feasibility
Project Scheduling and Budgeting

Lower row: Jennifer Doublet, Stephanie Erickson, Cecilia Recendez, Angela Brooks, Linda Jassim
Meddle row: Katrin Terstegen, Lori Bruns Selna, Vanessa Hardy, Stephanie Ragle, Lawrence Scarpa, Ching Luk
Upper row: Silke Clemens, Christopher Ghatak, Justin Patwin, Joshua Ashcroft, Royce Sciortino, Gwynne Pugh, Heather Duncan(missing), Dr. John Ingersoll(missing), Steve Kodama(missing)

Site Evaluation and Economic Analysis
Design/Build Project Delivery
Program and Construction Management
Furniture Design
Graphic Design

Clients

INSTITUTIONAL/GOVERNMENTAL

Santa Monica Museum of Art
Getty Conservation Institute
City of Santa Monica
Community Redevelopment Commission, Los Angeles County

Housing Authority of the County of Santa Clara
City of Upland
City of Palmdale
City of Carson
City of Los Angeles
County of Los Angeles Department of Parks and Recreation
City of Santa Monica Department of Parks and Recreation
Mountains Recreation and Conservation Authority
Housing Authority of the City of Alameda
City of Oakland Housing Authority
The Institute for Human and Social Development, Inc.

EDUCATIONAL

Los Angeles Community College District
Los Angeles Trade Technical College
West Los Angeles College
Santa Monica College
Claremont College
University of California at Berkeley
Stanford University
City College of San Francisco
San Francisco State University
Oakland Unified School District
San Francisco Unified School District
San Mateo Union High School District
Alum Rock Union Elementary School District
Jefferson High School District
Hebrew Academy of San Francisco

HOUSING

Prototypes
Livable Places, Inc.
Community Corporation of Santa Monica
Venice Community Housing Corporation
Charities Housing Development Corporation
Tenderloin Neighborhood Development Corporation
Mercy Housing California
San Francisco Housing Development Corp.
Ecumenical Association for Housing
Fine Arts Limited Partnership
East Bay Asian Local Development Corp.
Eden Housing
Mid-Peninsula Housing Coalition
Oakland Community Housing, Inc
Santa Cruz Community Corporation
Southern California Housing Corporation
Jewish Family and Children's Services
Citizen's Housing Corp.
Christian Church Homes of Northern California
BRIDGE Housing Corp.
Japanese American Religious Federation
Lexington Communities
Chinese Community Housing Corp.
Rubicon Programs, Inc.
Gramercy Group Homes
Progress Foundation
Allen Temple Arms Housing & Economic Development Corp.
Resources for Community Development
Berkeley Oakland Support Services
Stepping Out Housing, Inc.
Marin Association for Retarded Citizens (M.A.R.C.)
Cedars Development, Inc.
Peninsula Association for Retarded Children and Adults (P.A.R.C.A)
Rural Communities Housing Development Corp.
Serra Homes, Inc.
Community Housing Developers
Rubicon Program, Inc.

PARTIAL CLIENT LIST

Housing Authority of the County of Los Angeles
Vermont Slauson Economic Development Corporation
Community Development Commission of Los Angeles County
City of Santa Monica
Southern California Housing Development Corporation
Century Housing Freeway Program
Hill-n-Dale Child Care
Savo Island Cooperative Homes
Ecumenical Association for Housing
Marin Association for Retarded Citizens
Oakland Community Housing, Inc.
Mid-Peninsula Housing Coalition, Palo Alto, California
Church of All Nations
Cupertino Housing for the Disabled
Little Zion Missionary Baptist Church
Archdiocese of San Francisco
Chinese Community Housing Cooperation
Peninsula Association for Retarded Children and Adults
YMCA
The Salvation Army
Progress Foundation
Japanese Religious Federation
The Getty Conservation Institute
Cultural Affairs Department, City of Los Angeles
Santa Monica Museum of Art
El Pueblo de Los Angeles Cultural Museum
Bergamot Station Arts Complex
Kew Botanical Gardens London
Community Development Commission, County of Los Angeles
City of Santa Monica
City of Carson
City of Palmdale
City of Alhambra
Metropolitan Transit Authority
County of Los Angeles Department of Parks and Recreation
Mountains Recreation and Conservation Authority
City of Santa Monica Department of Parks and

Recreation
City of Upland Community Development Department
Los Angeles Trade Technical College
West Los Angeles College
Santa Monica College
Claremont College
University of California Irvine
University of California Berkeley
University of California Santa Barbara
Stanford University
Lawrence Berkeley Laboratories
The Institute for Human and Social Development
Community Corporation of Santa Monica
United States Department of Housing and Urban Development
Head Start Child Care Centers
Prototypes
Lee and Associates
Beitler
Coca-Cola Bottling Company
Cott Industries

Publications

2004
Ainge Magazine, "light Puzzle"vol,111
Los Angeles Business Journal, June 28, 2004
The Other Office-Creative Workplace,Birkhavser
Santa Monica Mirror, Sept.15-21
Santa Monica Mirror, Aug.18-24
Santa Monica Mirror, June.9-14
SPACE: International Interior Design Volume 1
HOW Magazine, June
Interior Design Magazine, June
Affordable Housing Finance, May
FRAME Magazine, March/April
Architectural Record, March
AIArchitect, January
Atlas of Contemporary World Architects, Phaidon Press Ltd

2003
Architectural Record, Record Interiors, September
FX Design, Business and Society, September
AIA J, National Journal of the American Institute of Architects, December
Abstract Magazine, November
Metal Architecture Design and Construction, Birkhauser DBZ Publications
Los Angeles Times, "Green Project Wins on its Own Terms", August 3
Architecture Now II, Taschen Publishing
Interior Design, January
Metropolis, March
Architectural Record, May
FRAME 30, January
ARCAca, September
ARCAca, February
DBZ Architecture, Winter issue, Germany
The Other Office, Peter Huiberts Publishing
Loud Paper, Vol. 4
LA Architect, Jan/Feb
Santa Monica Mirror, April 30
Hinge Magazine, June
BAUWELT, November, Germany
Atlas of Contemporary World Architects, Phaidon Press Ltd
Contemporary American Architecture Guidebook, Synectics Inc.
Interiors Now, The Images Publishing Group Pty Ltd
A+ D Home Front-New Developments in Housing (cover), John Wiley and Sons, Ltd.

2002
Architectural Review, November
Architectural Review, October
Interior Design, October
Santa Monica Mirror, October 23
LA Architect, July/August
The Art of Portable Architecture, Princeton Architectural Press
Dwell Magazine, August
Architectural Review, July
Interiors and Sources, March
LA Architect, Jan/Feb (cover)
DBZ Architecture "Glas 10", Summer issue Germany

AW Architektur + Wettbewerbe 191, September
The Hollywood Reporter, Fall Real Estate Special

2001
GA Houses No. 68

Grid Magazine, October
Buro DBZ ein Sonderheft der (cover)
ARCAca, 01.03 Summer
Architectural Review, August
Architecture Magazine, August
Los Angeles Times, June 26
Los Angeles Times, October 14
Lotus No. 107 Indoor Outdoor

Architectural Record, May

Architectural Review, March (cover)

Interiors Magazine, January

Architectural Record, January

LA Architect, May/June

Building Design and Construction, May

2000

Interior Design Magazine, March (cover)

LA Architect, November

LA Architect, June

Lotus, no. 105

Santa Monica Mirror, December 6

Monument, June

Metropolitan Home, September (cover)

1999

DBZ Architecture "Buro 99", Summer Issue, Germany

Domus, April

Interior Design Magazine, March

Journal of Architectural Education (JAE), September

FRAME, Vol #9, July/August

Architectural Record, January

Architecture Magazine, May

Des-Res Architecture, John Wiley and Sons, Ltd.

Context 3 Korea - Architecture Urbanism Landscape

1998

Architectural Record, December

Abitare, No. 378, November

Architectural Record, September

Interiors Magazine, September1998

Los Angeles Times Newspaper, June 23

Architecture + Urbanism (A+U), January

AD Tracing Architecture, John Wiley and Sons, Ltd.

DBZ Architecture "Buro 98", Summer Issue, Germany

Interior Design Magazine, May

The Hollywood Reporter, September 15-21

1997

Architectural Record, January

World Architecture, May

1996

Architecture Magazine, June

Architecture Magazine, March

1993

Progressive Architecture, November

1992

Progressive Architecture, October

Architectural Review, September

1991

GA House #31, Special Homes Issue (two houses),Japan

Progressive Architecture, March

Contributing Author, Contemporary Masterworks, St. James Press, London

1990

What is Post Modernism, Jencks, C. London, England

Los Angeles Times Newspaper, May

San Diego Constitution, October 10

Blueprint Magazine, London England

Architecture + Urbanism #229, Japan

Awards

2004

Recipient, National AIA Honor Award for Design Excellence

Exhibit, National Building Museum, Washington, DC, "Designing and American Asset - Affordable Housing"

Recipient, LA/AIA Honor Awards for design excellence (two awards)

Recipient, AIACC Honor Awards for design excellence (two awards)

Exhibit, LA 25, Los Angeles

Exhibit, National Building Museum, Washington, DC, "Liquid Stone"

The Architectural League of New York, "Emerging Voice" in architecture

Recipient, National AIA Honor Award for Design Excellence, CoOP Editorial

2003

Record Interiors, CoOP Editorial

Recipient, National AIA Honor Award for Design Excellence, Colorado Court

Recipient, Top Ten Green Building Award, National Award

Recipient, National AIA PIA Award for Design Excellence

Recipient, AIACC Honor Awards for Design Excellence, Colorado Court Recipient, Rudy Brunner Prize Silver Medal ($10,000.00)

Finalist, World Habitat Award

Recipient, National Endowment of the Arts Grant (Livable Places) $50.000.00

Recipient, LA/AIA Honor Awards for Design Excellence, CoOP Editorial

Recipient, LA/AIA Honor Awards for Design Excellence, Colorado Court

2002

Recipient, National AIA Honor Award for Design Excellence, XAP Corporation

Recipient, $50,000 National Endowment of the Arts Grant for Livable Places

Recipient, AIACC Honor Awards for Design Excellence, Bergamot Artist Lofts

Recipient, AIACC Honor Awards for Design Excellence, The Firm

Recipient, LA/AIA Honor Awards for Design Excellence, Youbet

Recipient, LA/AIA Honor Awards for Design Excellence, Davie Brown Entertainment

Recipient, $1,000,000.00 Operational Grant from Irvine Foundation for Livable Places

2001

Recipient, National AIA Honor Award for Design Excellence, Reactor Films

Recipient, LA/AIA Honor Awards for Design Excellence, XAP Corporation

Interiors Magazine "Project of The Year" Small Office Design, The Firm

Recipient, Westside Prize for Urban Design

Recipient, Los Angeles Business Council Design Award

Recipient, $500,000.00 Grant for Alternate Energy Research and Implentation

2000

Recipient, LA/AIA Honor Awards for Design Excellence, Reactor Films

Recipient, LA/AIA Honor Awards for Design Excellence, Bergamot Artist Lofts

Recipient, LA/AIA Honor Awards for Design Excellence, The Firm

Recipient, LA/AIA Honor Awards for Design Excellence, Click3xLA

Recipient, Westside Prize for Urban Design

1999

National Award for Excellence, Association of Collegiate Schools of Architecture (ACSA)

1997

Citation Award, Design Excellence in Architecture and Public Art awarded by the City of Los Angeles Department of Cultural Affairs

1996

Selected, Academy of Architecture Arts and Sciences, Top 39 Architects Under Age 39

Citation Award, Building Integrated Photovoltaics Design Competition.

1995

Recipient, AQMD Research Grant ($40,000.00) for Solar Design & Research

Finalist, "First Interstate International Affordable Housing (with Roger Sherman) Competition", ($25,000.00 Prize) South Central Los Angeles

1998

Winning Entry, Follansbee Steel International Design Competition

1985

Recipient, American Institute of Architects "Henry Adams Metal"

Chronology

图书在版编目(CIP)数据

皮尤+斯卡帕／蓝青主编，美国亚洲艺术与设计协作联盟(AADCU).
北京：中国建筑工业出版社，2005
（美国当代著名建筑设计师工作室报告）
ISBN 7-112-07385-5

Ⅰ.皮... Ⅱ.蓝... Ⅲ.建筑设计－作品集－美国－现代 Ⅳ.TU206

中国版本图书馆CIP数据核字(2005)第041315号

责任编辑：张建 黄居正

美国当代著名建筑设计师工作室报告
皮尤+斯卡帕

美国亚洲艺术与设计协作联盟(AADCU)
蓝青 主编
*
中国建筑工业出版社 出版、发行(北京西郊百万庄)
新 华 书 店 经 销
北京华联印刷有限公司印刷
*
开本：880×1230毫米 1/12 印张：16
2005年8月第一版   2005年8月第一次印刷
定价：**158.00**元
ISBN 7-112-07385-5
    (13339)

版权所有 翻印必究
如有印装质量问题，可寄本社退换
(邮政编码 100037)
本社网址：http://www.china-abp.com.cn
网上书店：http://www.china-building.com.cn